Classic Diesels of the South

A Railfan's Odyssey

by

J. Parker Lamb

TLC
PUBLISHING INC.

1997
TLC Publishing Inc.
1387 Winding Creek Lane
Lynchburg, Virginia • 24503-3776

Cover Painting: This painting specially commissioned for this volume by noted rail artist Andrew Harmantas, while not representing a specific J. Parker Lamb photograph is representative of the subject matter covered in this volume. Atlanta's Terminal Station displays a colorful array of first generation diesel power awaiting departure to points throughout the south.

Front Endsheet: A southbound Southern Railway local, working between Durham and Goldsboro, North Carolina, waits for ACL's northbound *Everglades* (No. 376) to finish its station stop at Selma. Youngster on bicycle seems to be in no hurry to ride away from this railroad action on a nice day in May 1962. Alco hood No. 10 is an RS-3 lettered for SR's Carolina and Northwestern (a collection of Carolina/Virginia shortlines).

Rear Endsheet: Seaboard GP-7 No. 1705 waits at the north end of the Raleigh, North Carolina passenger station as No. 7 the southbound *Sunland,* passes. The top headlight streak is from the oscillating light. Geep will pull in behind the train to take off cars in the August 1962 scene.

Dedication

This book is dedicated to the fond memories of my friendships with two North Carolinians, Wiley Bryan and Steve Hayworth. Over a period of nearly two decades I spent many enjoyable hours with each of them on train safaris throughout the South.

© Copyright 1997
TLC Publishing, Inc.

All rights reserved. No part of this book may be reproduced without written permission of the publisher or author, except for brief excerpts used in reviews, etc.

Library of Congress Catalog Number 95-62402
ISBN 1-883089-20-4

Layout and Design by
Kenneth L. Miller, Miller Design & Photography, Salem, Virginia

Printed by
Walsworth Publishing
Marceline Missouri

Contents

Foreword iii
Introduction vii
Chapter 1   Gulf States Beginning 1
Chapter 2   A Visitor's Perspective 25
Chapter 3   Carolina Ventures 49
Chapter 4   Riding Southeastern Lines 87
Chapter 5   End of an Era 97
Chapter 6   Epilogue 115

A winter sunset in 1984 glints off historic trackage which dates back to the original Mobile and Ohio mainline constructed through the Meridian-area in 1850. In the succeeding 146 years the rails have been owned by GM&O, ICG, MidSouth Rail Corporation, and currently by Kansas City Southern. Thus, one is forced to ponder the obvious question, "Will there be another owner in the near future?"

In June 1955 the northbound *Humming Bird* (No. 6), led by an E-8/E-7 lashup, picks up speed after crawling across the Biloxi Bay bridge in background.

Foreword

This volume represents an attempt to present in a coherent form some of the visual records of a railroad enthusiast whose photographic skills emerged during the early years of dieselization in the nation's southeastern corner which, for present purposes, extends westward to the Mississippi River and northward to the states of Kentucky and Virginia. What the reader will not find here is systematic coverage of each railroad in the region, but rather a photographic sampler which reflects the author's interest in first-generation diesel power on all the region's lines - both large and small. Interwoven around the descriptions of locomotives and operations are brief historical sketches of some key rail routes the author observed at close range during this period.

The commentary follows a chronological format, emphasizing various parts of the south where the author lived or visited. To understand how this photo coverage began, one must remember the importance of an "extended family" during the 1940's and 1950's. My parents were from large close-knit families, and we often visited my grandparents, uncles, aunts, and cousins located in cities and towns scattered throughout the south. These trips allowed me, as a dedicated train watcher in the beginning and later as a neophyte photographer, to gain an appreciation of the wide variety of rail operations and equipment on display in my "home territory." I soon realized that there were many who excelled at still photos and roster analysis, but far fewer who went into the countryside and attempted to create images of railroads doing their everyday jobs. As a mature photographer I came to realize that this visualization must include trains working in all kinds of weather and at all times of the day and night.

To those younger readers who were not around to watch this great metamorphosis of America's railroads, I hope these pages will give you an impression of what it was like. And to you who were there, I trust that these images will kindle some fond memories of the beginning of the diesel era.

J. Parker Lamb
Austin, Texas
March 1997

The village of Winfield (north of Birmingham) was a common meeting point for Frisco's *Kansas City-Florida Specials.* On an overcast afternoon in May 1955 the operator has train orders ready to hand to train No. 105, headed by an E-8 named *Gallant Fox* (right). Westbound counterpart No. 106 is led by *Pensive.*

On rare occasions the *Seminole* (No. 10) was run in two sections between the midwest and Florida. Here an IC E-7 (with green flags) and a CG E-8 (No. 812) lead a cut of reefers and head-end cars through a superelevated curve east of Columbus, Georgia (but still in Alabama) in February 1955.

Introduction

By the early 1950's, steam power in the southeast was employed only on those lines easily accessible to major shops and coal supplies, and by 1955 virtually all steam had disappeared. But why did dieselization proceed so rapidly in this region? The answer: it was an absolute necessity. Recall that most railroads of the south had been destroyed physically during the Civil War. After being painstakingly rebuilt, they had to endure a long series of economic recessions, financial panics, and later federal control during World War I, only to be devastated again during the Great Depression of the early 1930's. Less than a decade later, the four-year onslaught of World War II traffic levels stretched all railroads to the limit of their capabilities.

Thus southeastern railroads entered the post-war period with worn-down physical plant, worn-out rolling stock, roadbeds laid with light rail, and thousands of weight-restricted timber trestles and bridges. There was virtually no incentive, or feasibility, to invest in the new super power steam locomotives then being produced. Even a casual review will show that steam engines with four-wheeled trailing trucks, to support large, high-performance fireboxes, were "few and far between" down south. For example, both their tonnage levels and their torturous terrain required the Appalachian coal carriers (Clinchfield, C&O, N&W, and Virginian) to operate powerful steam engines

Central of Georgia passenger train No. 3, a Macon-Columbus-Birmingham local led by a 4-8-2, eases into the Opelika, Alabama station in the early morning of a July day in 1952. In the clear at right is southbound freight No. 34 behind a trio of hood units (two Geeps and an RS-3).

The *Seminole* (No. 10) speeds southward through the Alabama countryside north of Opelika, Alabama in February 1955. Both IC and CG motive power was run through between Chicago and Columbus, Georgia.

while, in contrast, neighboring L&N's largest engine was a modern 2-8-4. Otherwise, only ACL and RF&P, with large 4-8-4's, and Seaboard, with ex-B&O 2-6-6-4's, operated modern, large steam locomotives, although Central of Georgia and NC&StL operated smaller 4-8-4's and Norfolk Southern Railway had some vest pocket 2-8-4's.

What the new diesel-electric locomotive offered southern lines was light axle loadings, virtually no pounding of the rails, and the flexibility of increasing tractive effort merely by adding units. In short, it was an ideal match for the region's terrain, roadbed, and traffic patterns. And the economics relating to fuel and maintenance, in comparison with steam, merely solidified the diesel's superiority. Thus after World War II the south became a second home for diesel salesmen from all four builders of the period (Alco, Baldwin, EMD, and Fairbanks-Morse), and railroads of this region, in turn, became the nation's proving ground for widespread use of diesel locomotives which, having displayed their potential capabilities during limited wartime service, were still regarded by many mechanical officers, with some justification, as cantankerous and unreliable, and not to be trusted with the entire traffic load.

For those who wanted to observe the seemingly endless variety of early diesel models, the southeast was the place to be in the 1950's as Alco and Baldwin, assisted by wartime restrictions on diesel locomotive production and their reputation as "locomotive builders" rather than "diesel engine builders," made significant sales in the years before EMD began to flex its GM-based technical superiority in the marketplace. Early in this period Alco or EMD cab units were usually the primary power for mainline trains while switchers of many models proliferated. Later, road switchers which had been designed originally for yard work and local freight or passenger runs, graduated to general mainline service and, in a related development, many roads began to install MU-receptacles in the noses of A-units so that they could be placed anywhere in a locomotive consist, including "elephant-style" couplings.

This led to the photographically appealing "mix master" period where almost any combination of units, irrespective of builder or model, would be lashed together, assuming their control systems and gear ratios were compatible, which had not always been true in the beginning. Finally, in the latter stages of this first phase of dieselization one began to see extensive repowering of early Alco or Baldwin units, occasionally with newer Alco engines but more commonly with highly reliable General Motors diesels. These hybrid units were usually referred to as "halfbreeds" by fans.

Dieselization patterns of the various southeastern roads varied greatly. For example, GM&O went all-diesel with Alco products whereas the Norfolk Southern Railway chose Baldwins (except for a trio of GE 70-tonners). Bigger roads such as ACL, L&N, and the Southern Railway acquired large fleets of cab units (mainly EMD's F-series) whereas SAL was an early convert to Geeps for road power, often using older cab units as boosters. Just the reverse was true on the Central of Georgia where older F-units were generally in the lead and hood units were boosters. Seaboard, and to a lesser extent GM&O and L&N, were somewhat unusual because they used excess passenger cab units on freight trains near the end of their service lives. In contrast to those roads with large rosters of freight cab units, both Illinois Central and N&W stayed with steam until the B-B hood unit had become the dominant type of freight power.

Of the early diesel builders only Fairbanks-Morse failed to make much of a dent in the southeastern market. Southern Railway bought five burly, C-C trucked Trainmasters for its mountainous CNO&TP line from Cincinnati to Chattanooga, where they operated generally as boosters, along with sixteen B-B road switchers. In addition Central of Georgia purchased four yard units and five road switchers for use on locals. But F-M's best showing was on the Virginian Railway which relied exclusively on thirteen Trainmasters and forty B-B hood units to move tonnage on its non-electrified lines.

FP-7 No. 621, built in March 1951, has just received its third (and last) paint scheme and is working a northbound local through Tullahoma, Tennessee, where it waits for a meet on a hot July afternoon in 1967. As delivered it wore dark blue with cream stripes which later became solid blue and finally this high visibility scheme which included a yellow nose, gay body, and red heralds.

Early on a June morning in 1955, a pair of Alco DL-105's accelerate the *Gulf Coast Rebel* out of Meridian on the last leg of its overnight run from St. Louis to Mobile.
Right: GM&O FA-1s wait in the Meridian roundhouse between runs in May 1955.

Gulf States Beginning

1

Among my earliest recollections of train watching is standing on the porch of my grandparent's house, which sat on a knoll about a mile from the tracks, observing westbound trains, often behind double-headed 2-8-0's, accelerating to mainline speed on the Southern Railway's Alabama Great Southern line (AGS). The time was 1938 (I was not yet 5 years old) and the location was my birthplace, the tiny crossroads community of Boligee, Alabama where the AGS crossed Frisco's line to Pensacola. At that age I knew nothing about AGS or steam engine wheel arrangements, only that I could see clearly when there were two locomotives pulling together. Nor did I know that many of those trains which I watched with childlike curiosity were headed some 55 miles away to the yards in Meridian, Mississippi where my father chose to move his family in August of 1938.

In recent years I've often wondered if my later fascination with railroading was a result of growing up in Meridian or whether it would have occurred anyway. Obviously I will never know the answer, but I do know that this east Mississippi town, which owed its existence to the construction of the Mobile and Ohio Railroad in 1855, provided a constant stimulus for a curious lad to explore and learn about railroading during the hectic days of World War II. At its peak, wartime traffic through Meridian, a junction of three Class I railroads, was over 100 trains per day.

For me, mobility was the key to my early education in railroading. As soon as I learned to ride a bicycle at age ten, I was able to pedal a couple of miles to watch switch engines at work in the joint Illinois Central-Southern Railway yard which also included a coaling trestle, a large locomotive backshop, and a nearly semi-circular, concrete roundhouse. Only a mile away from this yard was the roundhouse and yard of the Gulf Mobile & Ohio Railroad while the town's resident shortline, the Meridian & Bigbee River Railroad, had its shop yet another mile distant. Thus, by the time I was in junior high school, which sat across the

street from the Southern Railway mainline, I was a regular visitor at all three areas, often making a two-hour round trip in the afternoons after school was dismissed.

In common with most other young rail enthusiasts, my first interest was building models (HO scale) and I devoured the details of various modelers magazines but in 1947 began to read *Railroad* and *Trains,* and to appreciate the work of early rail photographers such as Philip R. Hastings, Richard Steinheimer, and Robert Hale. I also noticed that there were very few photos of trains in the south and wondered why, with all the action I saw, there were not more fans recording these scenes for others to enjoy. But my first interest in photography was merely to take shots of equipment for modeling purposes. Thus my first photos, taken in the fall of 1949 with my mother's 1930's vintage folding Kodak which used 116-size film, were standard roster shots of locos and cars. Soon however, I joined the camera club at my high school in order to learn about darkroom techniques from my physics teacher. At his urging I, like many others at that time, bought a 35 mm Argus C-3 camera for personal use but, before my graduation in 1951, I had learned to operate the school's 4x5 Speed Graphic as well as a large Graflex (single lens reflex).

Although the north-south M&O route was Meridian's first railroad, two later rail lines (forming an east-west route) made the town into a strategic junction for transport of Confederate military personnel and supplies during the Civil War, and eventually led to Meridian's destruction by Union troops. The post-Bellum period saw construction in 1884 of yet another line. Conceived and promoted by a prominent Meridian attorney, the New Orleans and Northeastern Railroad connected Meridian with the Crescent City. Also during this period an English banking firm, Emile Erlanger and Co., was able to assemble one of the South's first multi-road systems known as the *Queen and Crescent Route* which stretched from Cincinnati (Queen City of the Ohio River) to Chattanooga on the Cincinnati New Orleans and Texas Pacific Railroad, then down through Birmingham to Meridian on the AGS, and finally into New Orleans over the NO&NE. During this time Meridian became a major junction for passenger service since the Erlanger interests also controlled the Alabama and Vicksburg Railroad as well as the Vicksburg Shreveport and Pacific Railroad which jointly operated a line from Meridian to Shreveport.

Control of the Q&C passed to the Southern Railway System in 1890 and in 1901 the SR acquired the nearly bankrupt M&O. Thus for almost twenty years Meridian was a key transfer point for sleeping car service between southern or western terminals such as New Orleans, Mobile, or Shreveport and eastern or northern terminals such as Atlanta, St. Louis, or Cincinnati. It was during the first years of the twentieth century that a four-tracked, multi-story passenger station was built (1905) and the joint yard for the AGS, NO&NE, and A&V was constructed, along with the NO&NE locomotive and car shop. Incidentally, remnants of the original passenger station have been incorporated recently into a modern multi-modal transportation center for Meridian.

Financial volatility preceding the stock market crash of 1929 led to the breakup of the Q&C System. One result was that the lines between Meridian and Shreveport came under control of the Yazoo and Mississippi Valley Railroad, an Illinois Central subsidiary. A short time later (1932) Southern control of the M&O ended as the shorter road sank into a receivership which would last until its merger with the Gulf Mobile and Northern in 1939 to form the GM&O Railroad. In the meantime another Meridian lawyer was responsible for developing two lines which ran east and west from Meridian. In an attempt to break the stranglehold of the Queen and Crescent system on Meridian's railroad service, the Meridian and Memphis Railroad constructed in 1913 a 32-mile line northwestward into Union to gain a connection with the GM&N mainline which ran between Mobile and Jackson, Tennessee. GM&N in fact would purchase

Top Left: A March 1955 view from the GM&O roundhouse in Meridian finds Alco FA-1 No. 702 on the turntable. At right is a former locomotive tender being used as a water car for a steam-driven wrecking crane.
Lower Left: The heaviest and most powerful diesel locomotive of the first generation was the Baldwin Centipede (DR-12-8-3000). SAL bought the first production unit (No. 4500) in December 1945 plus thirteen others. No. 4503 rests between runs at Columbus, Georgia in March 1954. At the time the huge but unreliable units had been demoted to branch line runs in Georgia and Florida.

The sun has already set on a hot June day in 1955 as the northbound *Gulf Coast Rebel* pauses at the Meridian station. After getting water for it train heat boilers, the two Alco DL-105 units will again highball on their overnight run to St. Louis.

the M&M in 1919, while in 1918 the Meridian and Bigbee River Railroad completed a 30-mile line eastward to Cromwell, Alabama and a connection with the Alabama Tennessee and Northern (later Frisco). The M&BR was extended across the Tombigbee River into Myrtlewood, its present terminus, in 1935 to connect with the L&N. A 1952 reorganization changed the name to Meridian and Bigbee Railroad.

During my senior year in high school I became more serious about railroad photography and evolved from roster shots exclusively into the action photo arena. Again the key was personal transportation. I had received my driver's license and was now able to commandeer the family car occasionally which allowed me to utilize choice photo locations on the various mainlines radiating from Meridian. Plying these routes was a staggering variety of motive power during the first half of the 1950's. The GM&O, which had become the nation's first Class I road to completely dieselize in 1949, used Alco FA/FB units on mainline freights and RS-2's on locals while, on its premier passenger train, the *Gulf Coast Rebel* (Mobile-St. Louis), one could find Alco DL-105's, PA-1's, and occasionally Baldwin "baby face" units. GM&O was also a big operator of gas-electric motor cars which it had inherited from both its predecessor lines. Until May 1952 it operated a daily train each way between Meridian and Jackson, Tennessee which consisted of a motor car (with baggage compartment in the rear) plus a trailing coach. In the Meridian yard GM&O used Alco S-2's and, for extended periods, would also assign the one-of-a-kind Ingalls 4-S road switcher to the yard.

In parallel with actions of the GM&O, the

GM&O's aging wooden roundhouse in Meridian always seemed somewhat inconsistent with the era of modern diesel power. This photo in April 1955 was taken from the cab window of another RS-2 waiting for its next road assignment.

In June 1954 the hump pusher at Norris Yard included TR-2 No. 2402, NW-2 No. 2275, and a slug built on the frame of a steam locomotive tender. Foreground trackage is a loco transfer line from arrival/departure yard to servicing area.

Southern Railway in 1949 was able to completely dieselize the NO&NE line south of Meridian which was plagued by light rail as well as numerous bridges and timber trestles. During this period Southern freights were led by solid lashups of EMD F-units (from FT's to F-7's) while Geeps held down locals, and both Alco S-2 and EMD switchers were used for yard work. And on rare occasions SR's only NW-5 unit would be assigned to the Meridian yard. Passenger trains were headed by FP-7's, E-7's, and E-8's with occasional appearances by Southern's rare DL-109/110 units. In contrast to Southern's dieselization program, both Illinois Central and M&BR went straight from steam to Geeps. The short line's diesels were more photogenic than IC power because of their brighter colors (dark blue with yellow trim), and hence received a large amount of my attention. The bland but utilitarian Illinois Central paint scheme had little appeal for me in those days and I rarely photographed IC freights around Meridian whereas the single IC passenger train (connecting with Shreveport) arrived and departed in the dead of night.

Graduation from high school in 1951 led to my relocation to Auburn, Alabama for college. The village of Auburn was located on the mainline of the West Point Route which ran between Montgomery and Atlanta. This railroad was composed of two separate companies which had their beginnings prior to the Civil War. The Montgomery Railroad was organized in 1834 to build from the Alabama capital to West Point, located on the Georgia side of the state line. The first 32 miles were completed within ten years but the entire line was not finished until 1851 after its name had been changed to Montgomery & West Point. In 1870 the name was changed to Western Railway of Alabama, even though it operated in the eastern two-thirds of the state. Meanwhile the Atlanta & LaGrange Railroad, chartered in 1847, opened a

The semaphore arm drops into its "stop" position as two West Point Route FP-7's accelerate the southbound *Crescent* away from the Opelika, Alabama station in March 1955. Train at left, waiting for the passenger train to clear, is a local freight between Opelika and Montgomery.

Atlanta to Montgomery local No. 31, led by a West Point Route FP-7, pauses at Auburn, Alabama on a steamy afternoon in June 1953.

On a cold day in December 1952, Columbus & Greenville green and orange Baldwin road switcher (DRS 6-4-1500) No. 605 clumps onto the turntable at the Columbus, Mississippi shops. Companion unit No. 601 is at right.

line from East Point (six miles from Atlanta) to West Point in 1854. Three years later it was renamed Atlanta & West Point. Both A&WP and W of A came under control of L&N/ACL in the 1890's after periods of control by the Central of Georgia Railroad and by the Georgia Railroad, which operated a connecting line from Atlanta to Augusta.

During my first year in college I "hit the books" in a big way and did very little photography although I watched with interest as the West Point's Ten-Wheelers, Pacifics and Mikes were replaced quickly by Geeps and FP-7's. Even though diesel units of the two roads were mixed freely and all displayed the diamond-shaped West Point herald on their noses, the specific railroad name appeared on the side of each unit. The primary color was dark blue but there was a light stripe, lengthwise along the middle of the cab or hood side. Curiously the stripe was white for Western units and silver for A&WP locos, a subtle difference which did not register in black and white photos unless the units were clean or there were two different engines coupled together. Most manifest freights on this line ran at night so I photographed mainly local freight and passenger trains as well as the road's premier run, the *Crescent* (originally *Crescent Limited*), a three-road streamlined train between the nation's capital and New Orleans. The West Point Route provided a 170-mile connection between the Southern

On a warm day in June 1952 the Atlanta-bound *Crescent*, headed by a pair of FP-7's, slows for the Central of Georgia crossing at the Opelika, Alabama station. Coaling tower and standpipe for steam locomotives will soon disappear.

Railway at Atlanta and the L&N at Montgomery.

Moving to eastern Alabama also gave me the opportunity to explore another Class I road which I had first encountered on trips to Birmingham in my early teens. The Central of Georgia Railroad mainline crossed the Western of Alabama at Opelika, six miles east of Auburn. The line between Birmingham and Columbus, Georgia was a part of the Central's Columbus Division and had been consolidated as the Columbus & Western Railroad in the 1880's. The 29 miles between Opelika and Columbus were built in 1885 by the Western of Alabama whereas the 55-mile segment from Opelika to Goodwater was constructed by the Savannah & Memphis in 1874. The 68-mile gap to Birmingham was closed in 1888 by the Columbus & Western which was then merged into the Savannah & Western. This latter road was absorbed into the Central of Georgia upon its formation in 1895.

During the first decade of the twentieth century both the Central of Georgia and Illinois Central were under control of the legendary Edward H. Harriman who used these roads, with an ACL connection, to establish service between Chicago and Florida. Thus the C of G line between Birmingham, Columbus, and Albany, Georgia (the ACL connection) carried extensive perishable and seaport traffic. Although IC control of the Central did not weather the Great Depression, these traffic patterns continued until recent years when alternate routes of the Norfolk

A pair of pre-World War II Alco HH900 switchers, in green with cream trim, work the Birmingham Southern yard in Bessemer, Alabama on a hot day in May 1954. The BS Railroad was a U. S. Steel road and unofficial visitors were generally not welcomed; I was careful to park my car at a "good spot" and step out quickly for this photo. Birmingham Southern owned eight of these units (Nos. 81-88) which were constructed in 1937.

In October 1954 SAL's *Silver Comet* prepares to depart Atlanta's Terminal Station on its final leg into Birmingham. Southern switcher No. 8200 is momentarily idle.

Southern Corporation made this line, with its low-clearance tunnels and steep grades, redundant as a through-route. Indeed, until the Central was absorbed into the Southern Railway in 1963, the Harriman era was still symbolized at Thomas yard in northwestern Birmingham where two, parallel classification yards were bounded on the north by shops of IC/C of G, and on the south by facilities of the Frisco, a one-time suitor for the Central.

During my years as an underclassman my parents did not allow me to have a car so I was forced to use a city bus whenever I went to Opelika to spend a day at the joint passenger station which guarded a busy crossing. The C of G's early diesel color scheme was one of my favorites, and certainly one of the South's most elaborate, primarily blue and gray with trim of orange and yellow (actually imitation gold). These colors would last until 1960 when an almost solid Pullman green scheme was introduced. Photographing the C of G was always interesting because of the variety of power.

A standard freight diesel lashup on the Albany-Birmingham segment usually included at least two (and often three) models. The road had bought ten "chicken wire" F-3's in 1947-48 and one of these was usually positioned at either end of a reversible consist. During the early 1950's Geeps or RS-3's were almost always used as booster units and seldom ran in the lead except on locals. Other than a Columbus-Birmingham local which disappeared in 1952, the only passenger service on this line was the joint IC-CG-ACL *Seminole* (Chicago-Jacksonville) which was led on alternate days by the Central's only two E-8's (Nos. 811 and 812). These units were usually run as a pair between Chicago and Columbus, so it was possible to photograph the 811 leading the train south before noon and then catch 812 heading north in mid-afternoon.

In 1954 I was able to buy a car as well as a German-made twin-lens reflex camera. My newfound freedom of movement allowed me to photograph most of the roads in Mississippi and Alabama during school vaca-

tion periods in 1954-55. In particular it was enjoyable to return to choice locations I had stumbled upon in my teen years but had never been able to exploit. I began submitting photos for publication in 1953 and, because of the encouragement of *Trains* editor David Morgan, began to realize that my efforts to explore southeastern railroads were of interest to a national audience. However, my days as a college student from Meridian, Mississippi were nearing an end.

On a warm August afternoon in 1957 the northbound *Gulf Coast Rebel* speeds toward Meridian behind veteran Alco No. 271. This scarred and weary unit was soon retired and the *Rebel* was discontinued in October 1958. GM&O's two early Alco passenger units (270 and 271) were classified as DL-105's when constructed in September 1940. They were the prototypes for the more famous DL-109 model which used the same carbody design.

An Albany, Georgia to Birmingham train grinds uphill through a long S-curve near Irondale, Alabama on what appears to be a double track section. In fact, for many years the CG and Southern operated their single-tracked lines parallel for six miles between Leeds and Irondale. This train was approaching the Weems crossovers which allowed CG trains to swing around the Norris yard complex on their route to downtown Birmingham.

Above: The lineup of classic EMD power from three railroads at Atlanta's Terminal Station on a blustery morning in October 1954 includes an E-7, FP-7, E-6, NW-2, and E-4.

Below: On a warm day in July 1952 the westbound *Crescent* (No. 38) clumps across the Central of Georgia diamond at the Opelika, Alabama station behind a pair of West Point Route FP-7's. Baggage carts are loaded with mail bags and ready for a quick transfer.

Central of Georgia hotshot No. 29 (Albany, Georgia to Birmingham) is powered by a seven-unit lashup of Geeps, RS-3's and "chicken-wire" F-3's (at each end) as it charges through a heavily forested area between Columbus, Georgia and Opelika, Alabama in June 1960. Somber Pullman green paint scheme was introduced in early 1960, replacing brighter colors of original CG diesels.

Left: On a July afternoon in 1954, a New Orleans-bound train pulls out of the Meridian yard, passing a concrete bridge pier which once carried tracks of the Meridian and Memphis Railroad over the mainline of the New Orleans and Northeastern. Later the M&M became part of the GM&N while the NO&NE came to be a component of the Southern Railway System. In the background is the junction of NO&NE and Alabama & Vicksburg (later Yazoo & Mississippi Valley).

On a warm day in May 1954 Birmingham-bound Central of Georgia freight No. 29, led by two F-3s spliced by an RS-2 and a Geep, heads into the siding at Smiths, Alabama (10 miles from Columbus, Georgia) to meet Albany-bound freight No. 34. As the train approached I asked if its headlight could be turned on, and the friendly crew radioed my request to the engineer. Earlier they had invited me into the cab when I walked up to ask how long they would be waiting.

On a February day in 1955 northbound Central of Georgia freight No. 29 grinds upgrade near Smiths, Alabama (near Columbus, Georgia) enroute to Birmingham and interchange with Frisco and Illinois Central at their joint yard in northeast Birmingham.

The southbound *Seminole* (No. 10) speeds through a cut north of Opelika, Alabama in February 1955. Motive power of both IC and CG was used between Chicago and Columbus, Georgia on this IC-CG-ACL train from Chicago to Jacksonville. E-8's 811 and 812 in this scene would later receive IC brown and orange paint scheme although still lettered for CG.

The joint yard in Meridian was a legacy of the Queen and Crescent Route formed in 1881. A common site was motive power of both roads waiting to go to work. A group of black IC Geeps edged into this early morning lineup of Southern Geeps and an FT set in August 1954.

On a cold November day in 1954, Southern Railway's *Birmingham Special* (No. 17) slows for its stop at Attalla, Alabama (between Chattanooga and Birmingham) behind an F-3/FP-7 combo. L&N's line from Birmingham to Anniston is in foreground while SR line was once part of the British-owned Alabama Great Southern.

The southbound *Seminole* (No. 10), a joint IC-CG-ACL train between Chicago-Birmingham-Jacksonville, pulls across tracks of the West Point Route into the Opelika, Alabama station in December 1954. Both IC and CG motive power was operated between Chicago and Columbus, Georgia.

On a cold day in December 1954 the eastbound *Kansas City-Florida Special* (No. 105), behind a pair of E-8's named for famous race horses, eases into the station at Tupelo, Mississippi. Crossing in the foreground is mainline of GM&O between Jackson, Tennessee and Meridian.

A battle scarred veteran FA awaits its next mainline assignment at the Meridian yard in August 1961.

At dawn on an August morning in 1952, northbound Central of Georgia freight No. 29 from Albany, Georgia pulls past the West Point Route crossing at the Opelika, Alabama station. Train will set out and pick up cars before continuing onto Birmingham. In these early days of diesel power only three units were needed for the relatively light train tonnages on this line.

An early morning scene in August 1958 at the Meridian passenger station finds the southbound *Gulf Coast Rebel* pulling out for Mobile behind a pair of DL-105's while a Birmingham-bound Southern Railway freight, led by a quartet of F-7's, crawls past on the outside tracks. Passenger train will have to wait for slower freight to clear crossover ahead.

Central of Georgia local No. 3 (Macon-Columbus-Birmingham) accelerates away from Opelika, Alabama early on a summer morning in July 1953 behind E-7 No. 806.

On a hot afternoon in July 1957, veteran E-6 No. 773 speeds through Ocean, Springs, Mississippi with the *Humming Bird* (No. 5) on its last lap into New Orleans.

A Visitor's Perspective

2

In late June 1955 my bride (of two weeks) and I packed our belongings in my 1950 Buick Special and a rented trailer, and began a 750-mile trek from Meridian to Dayton, Ohio where I would spend two years as a Lieutenant in the U. S. Air Force. My wife, whom I had known since junior high days and who was very familiar with my interest in railroad photography, was nevertheless taken aback when I pulled off the highway south of Nashville to grab a few shots of a GE 70-toner on an L&N work train which included an ancient steam-powered ditcher. I defended my actions as an extremely rare sighting which I simply had to record on film, but I sensed immediately that such an explanation was having very little impact.

The two years in Ohio were filled with much railroad exploration, a description of which must be deferred for now. Although Meridian was no longer "home," I was a frequent visitor there as well as in the coastal city of Biloxi where my wife's parents had settled in 1955. For many years there were at least two trips per year to visit family and relatives. During these times I usually managed to carve out a few half-day outings to "catch up" on railroad operations and equipment in the Meridian area and to learn more about railroading along the Gulf Coast.

The only mainline along the Mississippi coast belonged to the L&N's New Orleans and Mobile Subdivison (Montgomery, New Orleans & Pensacola Divison) and had been constructed in 1870 as the

In June 1955, while enroute from Meridian to Dayton, Ohio, I came across this scene on double track south of L&N's Radnor yard near Brentwood, Tennessee. A work train, powered by GE 70-ton switcher No. 98, was enlarging drainage ditches using an ancient steam-powered shovel which could shuttle on rails along its flat car support. No. 98 would later be fitted with an Alco 251B engine in 1966 and was eventually sold (1980) to Tropicana Corporation for switching duties in New Jersey.

The northbound *City of New Orleans* pulls out of the Jackson, Mississippi station on a frigid day in December 1955. The heavily trafficked mainline through the Mississippi capital was elevated in order to eliminate grade crossings.

New Orleans, Mobile and Chattanooga. A year later the name was changed to New Orleans, Mobile and Texas, and a decade after construction it was leased to, and later purchased by, the L&N. Because there were numerous stretches of this line within a mile or less of the sea front, the original builders had to construct six major crossings of bays and rivers along with dozens of shorter spans. The longest bridges were at Biloxi Bay (6500 feet) and St. Louis Bay (two miles). Indeed there were nine miles of bridge and trestle work on the 139-mile line. Symbolically this was where the L&N was forced to "walk on water." Not surprisingly the NO&M line has been virtually destroyed three times since World War II by major hurricanes and has suffered extensive damage on many other occasions.

Although in recent years this segment has emerged as a key CSX link in a southern transcontinental freight route between Florida ports and California, during the latter half of the 1950's this line handled far more passenger trains than freight. In addition to the fleet of traditional L&N trains between New Orleans and Louisville/Cincinnati (which included the *Pan American, Humming Bird,* and *Azelean*), there were two trains which operated through Atlanta via the West Point Route (*Crescent* and *Piedmont Limited*) and one Florida train (*Gulf Wind*). Moreover, there was a locally important commuter service between the Crescent City and Ocean Springs, Mississippi (90 miles) which was begun in the 1920's.

Motive power on the NO&M during this period was largely EMD cab units with E-6 and E-7 units being MU'ed with FP-7's on passenger runs while freights ran behind F-7 and FP-7 units with an occasional Geep as a booster. The ubiquitous FP-7's were also used on local freights along with Geeps. Because of the Automatic Train Stop system installed on the NO&M (until 1961) one found a relatively small number of properly equipped units constantly shuttling between Mobile and New Orleans. Fortunately, this process did not become boring because the diesel color scheme in those days was the original dark blue with yellow stripes and red trim (with black replacing blue on freight units), clearly one of the South's classiest looks.

One of my favorite spots along the Mississippi coast was in the town of Pascagoula where one could stand along an uncluttered water front and catch trains on a short deck-girder bridge and truss swing-span over the Pascagoula River. While waiting between approaching trains. As to why this was necessary one should remember that train radio was not yet in use and hence there were no scanners in those days. My main frustration with finding trains along the NO&M line was that the CTC signals were "approach lighted" which meant that, when one saw a signal light began to glow, only a few minutes would elapse before a headlight came into view.

On some of my twice-yearly trips between Dayton and Mississippi, I managed to spend a day at Central City, Kentucky which was the hub of steam power for

A rare snowfall on the Mississippi Gulf Coast in December 1963 provided an unusual setting for the northbound *Azalean* (No. 4) as it passed through Biloxi on its run to Mobile and Nashville behind an FP-7/E-7 lashup.

trains I found myself mesmerized by the ever busy sea gulls and the slow moving river boats which kept the bridge tender on his toes. Indeed I soon learned that one could usually use the closing of a normally open draw span in the same way as one would use CTC block signals, namely as a early warning device for IC's coal hauling Kentucky Division between Paducah and Louisville. On one occasion I decided to follow L&N's main line north from Knoxville to Corbin, Kentucky. This segment had always fascinated me because of its mountainous terrain as well as L&N's liberal use of Alco units in this area. I did manage to

The Southern Railway owned but a half-dozen Alco PA's which ran exclusively on the Memphis-Chattanooga-Bristol line during a relatively short operating career between 1953 and 1963. In August 1958 two of these sleek beauties pass on the outskirts of Memphis. Train No. 36 (left) is heading east while No. 35 is in the hole. Both trains are local runs.

find a few good locations for photos but alas, after the film was developed I found that my trusty reflex camera had suffered a fatal wound in the line of duty; one of the lens elements had been jolted out of alignment. Sadly, I've never had the opportunity to return to that area to get any more photos. On another trip I had planned a brief stopover in Oneida, Tennessee to inspect the Tennessee Railroad shops. However, on the day I arrived I found only dead engines (three steamers and three RS-1's) due to the year-end holidays. Moreover, the town was surrounded by a thick blanket of fog and drizzle which prevented anything but "record" shots.

Toward the end of my military tour, I decided to enter the University of Illinois at Urbana-Champaign in the fall of 1957 for graduate studies. With a two month break between my Air Force separation and the beginning of school, my wife and I spent almost six weeks visiting relatives in Alabama, Mississippi, Tennessee and Louisiana, introducing ourselves to the "other side of the family." Knowing that this brief interlude was probably my last open period for a while, I took time out for a number of trips to record steam powered southern shortlines in their twilight. But there were also opportunities to explore the hilly terrain around Birmingham for action on SAL, CG, SR, and Frisco. And in Memphis one afternoon I finally hit pay dirt in my longtime desire to photograph Southern's elusive Alco passenger units, which ran the length of Tennessee between Bristol and Memphis, when I happened upon a meet in which each train was led by a long-nosed PA-2.

My relocation to central Illinois in the fall of 1957 brought new opportunities to explore midwestern lines I had never seen, but it also brought me in contact with other like-minded railfans for the first time. From the perspective of the present era, it may seem incredible that one could roam around the south for seven years as I had, and never once encounter someone else who enjoyed following trains onto the main-

Framed by the truck and airhose of a bad order car on the shop track, GM&O FA No. 729 rests between runs at Meridian in May 1960. Note crease on far side of nose. This type of low angle shot required the photographer to sit on the ground.

lines for action shots, but it had happened. Of course, my lack of any connection to the sparse network of southeastern railfans was largely a result of my living in small towns such as Meridian and Auburn. Nevertheless, during my first year in Champaign I became friends with two other students, Bruce Meyer and Phil Weibler, both of whom eventually became professional railroaders. After graduation, Bruce went with EMD, rising from a field representative to an outstanding career as a design engineer while Phil gravitated to the operating side as an engineer, first with the Rock Island and later on the Chicago and North Western. While in Champaign I also chased trains around central Illinois with two other students who later gravitated to the publishing sector of avocational railroading, Jim Boyd and Harold Edmondson.

Although Bruce, Phil and I made numerous safaris throughout the midwest in the late 1950's in search of operating steam power, we also ventured as far northeast as Toronto and as far southeast as Bluefield, West Virginia. It was on a trip to West Virginia in the spring of 1958 that I saw my first Virginian opera-

A pair of yellow and black Trainmasters (with No. 54 leading) rumble across the New River with a Charleston, West Virginia to Princeton train in June 1958. This scene, at Deepwater Bridge (DB Tower) where VGN rails joined those of the New York Central, is a record of my only encounter with Virginian diesels, a circumstance I've always regretted.

tions. Although we were intent on shooting N&W steam power, we also caught glimpses of the VGN electrics operating eastward out of Princeton. Later that summer I made another trip to this area alone and, while driving eastward from Charleston along the Kanawha River, happened upon an eastbound VGN train led by a pair of yellow and black FM Trainmasters. Much to my surprise it met a New York Central local, behind a pair of EMD switchers, at Deepwater Bridge Junction just before crossing the river. Not until I consulted my *Official Guide* that night did I realize that NYC had a line into this area over which VGN had trackage rights into Charleston. Unfortunately, this chance sighting turned out to be the only diesel-powered Virginian train I would ever encounter.

On a trip north from Meridian to Champaign in 1959 I decided to check out GM&O's Islen shop complex in Jackson, Tennessee as well as the ex-NC&StL terminal at nearby Bruceton. Arriving at Islen in mid-afternoon I grabbed a few shots of Alco units in front of the mammoth, glass-walled shop building, but what really peaked my interest were the three "little" *Rebel* trainsets in the scrap line. The shovel-nosed power cars, with one B-truck powered by a 660-hp Alco engine, contained a baggage compartment behind the engine room and pulled from three to four lightweight, matching cars. Built in 1935 and 1937 by American Car and Foundry (ACF) in St. Louis, they inaugurated GM&N's *Rebel* service

A low winter sun highlights undergear of the southbound *Hummingbird* (No. 5) as it rumbles across the recently silvered East Pascagoula River bridge behind and E-7/E-6 combination on a warm afternoon in December 1961.

between Jackson, Tennessee and New Orleans, with a stub operation from Union to Mobile, and represented the South's first streamlined trains. I never had a chance to photograph them in action before their retirement in 1955 because they passed through central Mississippi at night. Moving on to Bruceton near dark, I found a well lighted locomotive servicing area that evening and made a few time exposures of ex-NC&StL units still in their original colors of red and yellow for yard engines or blue and gray for road units. Unfortunately, all NC identification had already been neatly covered by matching paint and replaced by an unobtrusive *L&N*.

By the early 1960's, operations in Meridian had changed to reflect the longer freights (often over 150 cars) being operated by both Southern and GM&O. On both roads, manifest trains in the 1945-50 period were handled easily by three 1500-hp units. However, as forty-foot cars were replaced by 50-footers and later with the introduction of piggyback flats and auto racks, these longer and heavier trains required from four to six units over even the relatively flat profiles of most southern lines. By the mid-1950's GM&O had modified its FA's with nose-mounted MU connections so it could run these engines anywhere in the consist, except nose-to-nose. On the other hand, SR usually ran its trains from Birmingham to New Orleans behind a traditional lashup of A-units at each end with from two to four B-units sandwiched between, not an especially photogenic power consist but effective. Although GM&O hardly ever mixed its RS-2's with FA/FB units on manifests, Southern would occasionally drop a GP-9 into the lashup as a booster unit between Birmingham and New Orleans.

In order to increase the average speed of its fast manifests south of Birmingham, Southern began in

Fifteen miles northeast of Birmingham a quartette of SAL Geeps crawl through a deep rock cut with an Atlanta-bound train in June 1960.

At dusk on a hot August day in 1964, Illinois Central passenger train No. 208, the *Northestern Limited*, pulls out of Shreveport on an overnight round trip to Meridian. In the background a pair of KCS E-8's await the departure of the northbound *Southern Belle*. At Meridian the IC train will be turned and return as No. 205, the *Southwestern Limited*, arriving back in Shreveport in the early morning hours. Illinois Central continued this service until 1968.

early 1960 to assemble hotshots with cabooses interspersed in the consist. For example, No. 153 (Cincinnati to New Orleans) might leave Norris yard in Birmingham with nearly 200 cars consisting of a cut for New Orleans followed by a caboose. Then came (sometimes) a cut for Hattiesburg and its caboose, and finally a Meridian cut and its caboose. At the first stop in Meridian, the local cut was detached quickly as a new crew boarded the caboose behind the Hattiesburg cut. The procedure was repeated in Hattiesburg, thus saving time in each yard stop. The result was much greater efficiency for the railroad but it was bad for my train scouting procedure which consisted of "cruising by" the Southern-IC yard at various times during the day to see if any trains were being readied for departure.

By the end of 1960 I could see the completion of my academic work, and begin to think about permanent employment. My choices were either industrial research labs or academic institutions. I finally chose the later path and, after mulling over the question of location, my wife and I decided to concentrate a job search in the Virginia-North Carolina area. In June 1960 I made a trip from Meridian to Raleigh, North Carolina for a job interview. On the first day I stopped by Opelika to see how much things had changed at the junction, and later in Newnan, Georgia came upon an Atlanta-bound CG local behind a trio of RS-3's, two of which displayed Central's Pullman green scheme. The next day found me south of Spartanburg, South Carolina where I was lucky enough to catch a southbound Piedmont and Northern train powered by a

Two sets of Alco cab units, displaying the battle scars from more than 15 years of mainline work, await their next runs at the Meridian yard on a foggy night in August 1961.

pair of black and yellow RS-2's. My immediate goal was to stop over in Spartanburg and explore the interchange process between the Clinchfield Railroad and the ACL subsidiary Charleston and Western Carolina whose mainline ran from Spartanburg through Augusta to the coastal town of Port Royal. These two roads combined to operate a 500-mile link in the midwest-Florida route between ACL's mainline at Yemassee (near Charleston) and the C&O at Elkhorn City, Kentucky. This three-road routing competed for midwest traffic with Southern's Jacksonville-Atlanta-Cincinnati lines as well as L&N's Atlanta-Knoxville-Cincinnati segment.

The following day I was able to catch a few shots of Chinchfield's gray and yellow F's which were used in five and six unit consists. Later, heading south from Spartanburg I hit the jackpot when I overtook a southbound C&WC train headed by a pair of faded purple and white ACL F-7's and two black Geeps. This would be one of my last encounters with the distinctive purple colors made famous by longtime ACL President Champion McDowell Davis. The next day in Raleigh I would make my initial acquaintance with another classic paint scheme, Norfolk Southern's red and yellow Baldwins.

GM&O train No. 32 from New Orleans to Jackson, Tennessee and St. Louis, crawls upgrade through Red Cut east of Union on a frigid morning in December 1960 after a freak snow fall has covered central Mississippi.

A warm day in September 1954 finds NC&StL GP-7 No. 753, working between Bruceton and Memphis, resting at the station in Jackson, Tennessee while the crew has lunch. Note lack of MU connections.

In June 1954 a Geep-led Frisco local heads out of Boligee, Alabama on a run from Alicevlle to Demopolis. Boligee, where the Alabama Great Southern line crossed the Frisco, was my birthplace, but this is the only train I have ever photographed there.

In December 1954 a westbound Frisco freight coils through an S-curve near New Albany, Mississippi (80 miles east of Memphis) behind alternating Alco and EMD cab units, led by FA-1 No. 5207.

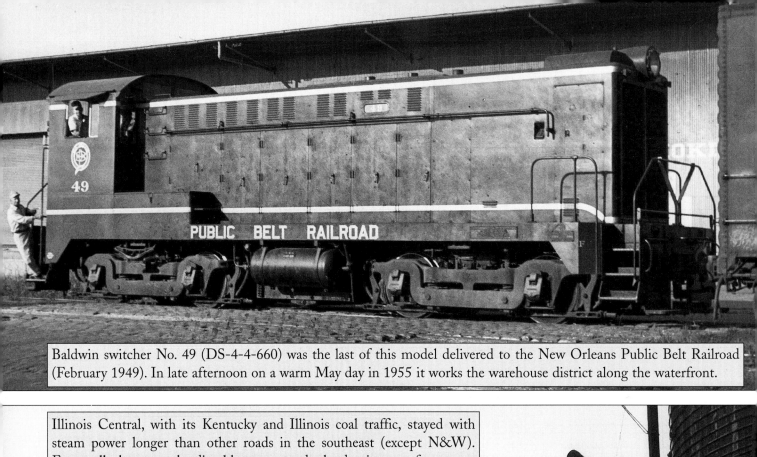

Baldwin switcher No. 49 (DS-4-4-660) was the last of this model delivered to the New Orleans Public Belt Railroad (February 1949). In late afternoon on a warm May day in 1955 it works the warehouse district along the waterfront.

Illinois Central, with its Kentucky and Illinois coal traffic, stayed with steam power longer than other roads in the southeast (except N&W). Eventually, however, the diesel began to erode the dominance of steam on both of these holdouts. Usually dieselization began with yard engines and IC was no exception. In December 1954 I came across this scene in Jackson, Mississippi which symbolizes the changing era of motive power.

A brace of E-7's lead the New Orleans-bound Hummingbird (No. 5) through the drawspan over the East Pascagoula River in July 1963.

The Ingalls Shipbuilding Corporation of Pascagoula, Mississippi attempted to break into the locomotive market during the post-World War II period, producing a prototype road switcher (Model 4S) in 1946. The unit was eventually sold to the GM&O but no additional orders were ever forthcoming. Thus No. 1900 became a somewhat cantankerous orphan during its 20 year service life. It seldom ventured north of Meridian (ex-M&O lines) or nearby Union, Mississippi (ex-GM&N lines), and worked only in yard switching service due to a lack of MU-capability. These two scenes at Meridian in July 1960 may be the only photos ever made which show the top of this locomotive, a rare link in the chain of development of diesel power on American railroads.

Columbus & Greenville Whitcomb No. 42 (65 ton, 650 hp) was built for the U.S. Army in July 1944. However, in April 1948, it was rebuilt by Whitcomb and sold to the C&G. The little diesel spent virtually its entire service life in switching duties around Greenville, Mississippi where this photo was taken in August 1957.

Above: The last day of 1958 finds an L&N freight leaving Attalla, Alabama for a run to Birmingham behind a pair of ubiquitous F-7's, the first of which carried an experimental color scheme featuring the gray body (later adopted) with triple diamond herald.

A warm evening in September 1959 finds a set of F-7's and two Geeps at the ex-NC&StL shops in Bruceton, Tennessee. All of these units were ex-NC power which was gradually receiving the black L&N scheme during the months following the merger.

Left: Classic cab units pause for a portrait at Gulfport, Mississippi in August 1958. The northbound *Azalean* (No. 4), led by FP-7 No. 651 and an E-7, has made a quick stop at the Gulfport station. Meanwhile, a pair of E-6's power a troop train which is loading Army reservists who have been in training at Camp Shelby near Hattiesburg.

In August 1958 the operator at Biloxi, Mississippi prepares to "hoop up" a set of orders to the engineer and conductor of eastbound local No. 74 (led by an FP-7) headed for Mobile. CTC was installed on the New Orleans-Mobile segment in 1961.

With the Spanish architecture of Birmingham's picturesque passenger station as a backdrop, classic diesels of Southern and Seaboard await their next assignments in June 1960. SAL E-4 is headed to Richmond with the *Silver Comet* while the SR E-8 will lead a section of the *Kansas City-Florida Special* (No. 8) toward Atlanta and Washington.

Right: GM&O freight No. 31 leaves the Meridian yard on a run to Mobile in August 1962 behind a lashup of five Alco cabs. The fifth unit was probably necessitated by the long cut of large steel pipe shipped from Birmingham mills. Mileage marker on bridge (at bottom of photo) signifies distance to Mobile.

A warm night in August 1962 finds SAL E-7 No. 3030 waiting for the higball in Raleigh while leading the southbound *Sunland* (No. 7).

Carolina Ventures

3

As I had done ten years earlier when I entered college for the first time, I laid aside the camera and attempted to concentrate on my work during the first few months after moving to Raleigh in January 1961. However, this abstinence was quite difficult since the North Carolina State University campus may have been the only one in the nation to be bisected by a two-railroad (SR-SAL) double-tracked mainline. Indeed my wife often accused me of letting that feature influence my job decision, and I have always maintained that it was "just luck." (What do *you* think?)

I stuck to my resolve to emphasize professional activities for almost three months until one afternoon in April when there was a major derailment in downtown Raleigh. A northbound SAL passenger local (Hamlet to Richmond), after passing through the campus area on a descending grade, failed to negotiate a sharp curve at the bottom of the hill. The train tipped over and slid on its side for a hundred feet, the locomotive (an E-7) coming to rest with its nose pierced by a large I-beam in a nearby storage yard. Miraculously there were no fatalities although the steel beam missed by inches crushing the engineer. Of all the railroad wrecks I have encountered, this one was the most easily accessible. There were street overpasses on two sides which allowed spectators to view the cleanup operations in an amphitheater-like setting, unencumbered by security personnel. While taking photos of the two steam-powered wreckers, I fell into conversation with another photographer. This chance meeting led to my lifetime friendship with Wiley Bryan.

Wiley, a native of nearby Durham, had caught on as a fireman with the Seaboard during the early part of World War II when many older workers went to battle. However, his railroad career turned out to be a brief one when he too had to enter military service later in the campaign. After the war he became a real estate appraiser for a mortgage banking firm and traveled throughout the state inspecting commercial property. Like me, he always carried his camera on trips and enjoyed the exploration and discovery of choice locations for mainline photography. With his thorough knowledge of the region we made numerous one-day trips, exploring segments of the SAL between

Led by E-6 No. 507, ACL's northbound *Everglades* (No. 376) is about to hit the crossing diamond in the eastern North Carolina village of Selma on a nice morning in May 1962. Waiting for clearance is a Southern Railway local working between Durham and Goldsboro. The lead RS-3 (No. 10) is lettered for Southern subsidiary Carolina and Northwestern.

In June 1960 a pair of orange and yellow Baldwin road switchers rumble eastward out of Raleigh on an overnight run to tidewater with Norfolk Southern Railway train No. 64. Despite the similarity of their carbodies, lead unit is an AS-616 while second engine is a DRS-6-4-1500.

Hamlet and Richmond. We also frequented the Coast Line yard in Rocky Mount which was the first classification point south of Richmond, just as Hamlet yard was on the SAL.

The 258-mile stretch of trackage from Hamlet to Richmond, part of Seaboard's Virginia Division, was the capstone of the entire SAL Railroad. South of Hamlet SAL trains could arrive or depart on five different routes (serving Wilmington or Bostic in North Carolina, Charleston or Columbia in South Carolina, and Atlanta-Birmingham). The second mainline segment of the Virginia Division was the line between Norlina (60 miles north of Raleigh) and Portsmouth, Virginia. It was along this tidewater route that the name Seaboard Air Line was inaugurated in 1873 to represent the consolidation of two pre-Civil War lines, Seaboard & Roanoke and Raleigh & Gaston. The northern-most sixty miles of the Virginia Division (north of Alberta, Virginia) was somewhat historic as

the first installation of CTC in the south. It occurred early in the Second World War when the SAL, then in receeivership and faced with a staggering traffic density, had to choose between building a second mainline (*a la* ACL) or going electronic. Eventually, the Virginia Division dispatcher's office near the SAL passenger station in Raleigh would house three CTC panels covering all mainlines.

My residence in Raleigh was an unexpectedly short twenty-four months, but it represented the most intensive period of railroad photography in my entire life. I now had both the time and finances to indulge myself in a region blanketed with a web of busy railroads. In 1961 Raleigh was, as Meridian had been a decade earlier, a train watchers paradise in terms of variety of motive power. For example, Seaboard used E-4, E-6, E-7, and E-8 units on its fourteen daily passenger trains through the Tar Heel capital while freights were hauled by passenger units mixed with

On April 6, 1961 northbound SAL mail and express train No. 4 derailed on a curve (between the two Geeps) due to excessive speed. The location was at the bottom of a long grade which ended in downtown Raleigh. After the train derailed, the lead unit, E-7 No. 3027, slid on its side into a storage yard, its nose being pierced by a large I-beam. Here we see southbound freight No. 75, led by a GP-9 and two FT A-B sets, passing the cleanup operations. Two SAL steam -powered wreckers are dragging the E-unit back to the tracks where it was hoisted and remounted on its trucks (one truck is in center-background). The accident happened in mid-day and by sunset the wrecked unit had been rerailed. By dark it was in tow back to the shops at Hamlet, later to be traded in on an E-8.

FT's, F-3's, and plenty of Geeps (both 7's and 9's). And occasionally one would see F-7's or E-8's from the Richmond, Fredericksburg & Potomac on run-through freights between Potomac and Hamlet yards.

Although SAL was the major player on the Raleigh railroad scene, one could find a significant amount of activity on the Baldwin-powered Norfolk Southern Railway whose meandering 383-mile main stem between Norfolk and Charlotte passed through the Tar Heel capital. The NS yard classified not only manifest freights, but was also the hub of operations for two busy branch lines which served Fayetteville (and Ft. Bragg) to the south and Durham to the north. Raleigh was also traversed by the Southern Railway line between Greensboro and Goldsboro from which its subsidiary Atlantic & East Carolina served the coastal community of Morehead City. At that time SR also tapped the Norfolk area via trackage

When this scene was recorded at Richmond's Acca yard in October 1961, the Atlantic Coast Line was in the process of repainting its units from the famous purple and white to a more somber black scheme. FP-7 No. 875 exhibited the original scheme while F-3 No. 341 was adorned with the newer colors. Longtime ACL President Champion McDowell (Champ) Davis was partial to purple but, after his retirement, the "black brigade" took over.

rights over the Coast Line between Selma (28 miles east of Raleigh) and Pinners Point yard. Southern's line paralleled the Seaboard mainline for eight miles between Cary and Raleigh, and the two roads operated this section as a double track into downtown Raleigh.

Occasionally Wiley and I would head south from Raleigh and go all the way to Hamlet for a day of photography, but more often we would stop at Apex (14 miles) and Aberdeen. The village of Apex was the primary SAL interchange point and operational hub for the Durham & Southern Railroad. Tapping the Durham industrial complex, D&S's major traffic consisted of southbound loads to SAL and bidirectional interchange with the ACL at Dunn, the southern terminus. Incidentally, SAL could take northbound loads out of Durham on its own branch from the mainline at Henderson. The D&S maintained a small amount of interchange with the NS at Varina, and carried substantial loadings of sand and gravel from quarries near Dunn to a large Durham construction company, which just happened to own a controlling interest in the railroad. At Apex the D&S had switching and interchange tracks on both sides of the SAL mainline and, in late afternoon, usually ran a northbound train to Durham and later tied up power for the Apex-Dunn turn run. When I arrived in North Carolina in 1961 the D&S was powered by a trio of black and white Baldwin RS-12's but in 1962 purchased a Soo Line DRS 4-4-1500. Aside from its rare locomotives the most distinctive rolling stock on the D&S was its fleet of homemade cabooses which featured electric

Coast Line's *Havana Special* (No. 75) pulls into a sparsely populated station at Selma, North Carolina on a nice May morning in 1962. Tracks in foreground belong the Southern Railway line between Greensboro and Goldsboro.

generators powered by small, roof-mounted windmills.

The small town of Aberdeen, 72 miles south of Raleigh, straddled the Seaboard mainline but was also home to the 45-mile long Aberdeen & Rockfish, and the terminus of a 33-mile Norfolk Southern branch from the mainline at Star. It was not unusual to see locomotives of all three railroads switching interchange cars at the same time, producing a kaleidoscope of color schemes, from the light blue and silver of A&R, to red and yellow of NS, to brown and yellow of SAL. The A&R, operating at that time with a 1947-vintage F-3A and a GP-7, dispatched two trains eastward from Aberdeen each morning (except Sunday). The first run worked the eastern end of the line around Fayetteville and the Ft. Bragg military reservation while the second went as far as Raeford (19 miles). What was just as interesting from a train watcher's perspective was the local topography. Aberdeen sat in a valley so that every train leaving

On a foggy day in April 1962 a pair of Tennessee Central RS-3's pump air at Nashville yard before heading east with a local.

While waiting for their locomotive to be released, two Norfolk Southern crewmen take a break at the engine servicing facility in Raleigh. Baldwins 1606 and 1611 were both AS-616 models.

town in any direction had to climb a hill. For example, northbound SAL trains faced a steady climb of four miles into Southern Pines, and the line was quite photogenic with a number of reverse curves. Early on, Wiley showed me one of his favorite spots, a highway overpass south of Southern Pines where one could get great shots in both directions.

When Wiley and I went north out of Raleigh we usually worked our way along the SAL to the Norlina junction and then occasionally on northward to Bracey, Virginia where there was a deep cut spanned by a highway overpass. On our way to Norlina we often stopped and waited for trains to cross the 500-foot long, 70-foot tall steel bridge over the Tar River north of Franklinton. If we were driving directly to Richmond for SAL and RF&P action, we usually left around 5:30 am since Wiley made a point of stopping at SAL's Petersburg station which, he claimed, had the best pancakes and ham breakfast he had ever eaten, quite a compliment from a man who spent most of his time traveling to cities and towns throughout North Carolina and Virginia.

One of the unique features of the SAL facilities in Raleigh was the close proximity of its two-tracked passenger station to the loco servicing and storage area (1000 feet to the south) and to the flat switching classification yard whose ladder tracks were directly to the west of the platforms. Thus, after dark there was ample illumination from light towers associated with these facilities to support night shots which required very little flash fill in. Inasmuch as my apartment was only a ten-minute drive from the station, on many evenings I would drop by after dinner to check out potential subjects such as locos resting between runs,

Piedmont & Northern's early history as an electric powered line is evidenced by a catenary support tower on the high bridge near Taylor, South Carolina. Two RS-2's power a southbound train from Spartanburg to Greenville in June 1960.

freight trains awaiting departure, or passenger trains making brief stops. As a result my success in making night photos in Raleigh exceeded my wildest expectations.

In April 1962 I drove to Nashville for a professional meeting and, along the way discovered not only the beautiful Blue Ridge scenery along the Southern Railway between Asheville and Knoxville but, west of Harriman, Tennessee, an entirely new (to me) railroad which has been exposed only rarely to the national railfan community. This mysterious line was the Tennessee Central whose 284 miles were operated as two separate railroads with Nashville as the hub. To the northwest of the Tennessee capital the 84-mile stretch to Hopkinsville, Kentucky served the important military post at Ft. Campbell while the 166-mile mainline east of Nashville was a true mountain railroad which interchanged bridge traffic with the SR and L&N on its east end, and served many coal mines and rock quarries along its route.

On my first trip I was able to pace No. 84, the daily eastbound run from Nashville to Emory Gap, and also grab some shots around TC's shops and yard complex which lay alongside the Cumberland River east of downtown Nashville. While obtaining permission to tour the shops from TC's general manager I inquired about the possibility of a cab ride and, to my surprise, he readily agreed. (See next chapter for a trip summary.) In early 1962 TC was primarily a first generation Alco road with a roster which included an S-1, nine RS-3's, and two FA's. These units were supplemented with three Baldwin switchers. Later (in 1962-63) the road would acquire a group of low hood Alco road switchers.

Conceived as an east-west railroad in a region where, in later years, the dominant traffic patterns would be north-south, the TC saw more bad times than good throughout its 84-year existence. Chartered as the Nashville & Knoxville in 1884, it was completed from Nashville to Monterey in 1894. In the meantime Tennessee lawyer and entrepreneur Jerre Baxter had chartered the Tennessee Central *Railroad* in 1893 to link Knoxville and Nashville with the Tennessee River in the western part of the state. He soon gained control of the N&K and envisioned it as the center segment of the Tennessee Central route. By 1900 a newly organized Tennessee Central *Railway* line reached Emory Gap and a connection with the CNO&TP (Southern Railway System). The Hopkinsville line was completed in February 1904, shortly before Baxter's death which led to a five-year receivership. Then followed a generally prosperous period of thirty years, leading to the lean times of the Depression era. After running its steam power "into the ground" handling the traffic burdens of World War II, TC dieselized in 1948 and continued to do well during the 1950's due to coal loadings for Tennessee Valley Authority (TVA) power plants. But traffic levels declined during the 1960's and a bankrupt TC was finally carved up by its surrounding Class I competitors in 1968. Illinois Central claimed the Hopkinsville line, L&N the central part around Nashville, and Southern the eastern thirty miles to Crossville. Eventually some of these segments spawned even smaller shortlines.

The highlight of my stay in Raleigh was the opportunity to ride freight trains over the entire mainline of the Virginia Division in preparation for a *Trains* article (May 1963). With clearance from the SAL Public Relations Department I was allowed free reign to roam through Hamlet yard and diesel shop as well as the dispatcher's office in Raleigh. Thus my knowledge of the inner workings of the Seaboard was far more thorough than for any Class I railroad I have ever

The Durham & Southern's homemade wooden cabooses with their propeller-driven electric generators were among the most unusual pieces of rolling stock I have ever encountered in nearly a half-century of train watching. Here are cars X83 and X84 resting between runs at the Apex, North Carolina yard in June 1962.

encountered. A summary of these SAL trips is presented in the next Chapter.

Despite my successes in, and enjoyment of, these railroad related activities in the Carolinas and Virginia, it was clear after a year and a half that another move was imminent. The professional opportunities I had envisioned when I joined the faculty of N. C. State University were not developing. Thus in the fall of 1962, after considering a number of possibilities my wife and I decided, with deep and mixed emotions, to *Go West*. For the first time we would be living west of the Mississippi River, in the capital of the Lone Star state.

Right: Slant-nosed E-4 No. 3002 was in the first order of SAL passenger diesels in late 1938. Here is the veteran unit carrying green flags at the Raleigh station in November 1962 while heading the southbound *Sunland* (First No. 7).

Although its road switchers could be easily operated in either direction, Norfolk Southern liked to keep the long hood forward whenever possible. Here we see DRS-6-4-1500 No. 1507 (built March 1948) getting a spin on the turntable at the Raleigh shops in August 1962. Note the airhose connection to motor which drives turntable.

Norfolk Southern's three GE 70-ton switchers (built June 1948) were the heaviest power allowed on the road's Bayboro, North Carolina branch. Here a pair work eastward toward New Bern on a toasty day in August 1962. Note chains and push poles hanging on locos, often a necessity on backwoods branches.

Right: Framed momentarily in a boxcar door, Norfolk Southern Baldwin switcher No. 663 is busy classifying trains in the Raleigh yard in May 1961.

A trio of purple Coast Line Geeps and a black F-7 coast downhill near Woodruf, South Carolina with westbound merchandiser No. 97 in June 1960. At Spartanburg, ACL-subsidiary Charleston & Western Carolina will hand the entire train over to Clinchfield for delivery to the C&O at Elkhorn City, Kentucky.

Led by a pair of purple Coast Line F's and two black Geeps, eastbound merchandiser No. 92 climbs a grade near Woodruf, South Carolina in June 1960. At Spartanburg the Clinchfield Railroad handed the train off to ACL-subsidiary Charleston & Western Carolina for delivery to the ACL mainline at Yemassee. These were the last purple-clad cab units I ever saw in revenue service.

Trailed by a short caboose with a baggage compartment, a westbound Aberdeen and Rockfish freight eases into the yard at Aberdeen, North Carolina in June 1962. Veteran F-3 No. 200 waits for clearance to make a reverse move into the engine terminal.

An eastbound C&O merchandiser heads along the James River line in downtown Richmond behind five GP-9's. A bascule lift span frames the train, a few boats and a solitary fisherman on a cool morning in October 1961. This location is near the famed tri-level crossing of SAL-SR-C&O at Main Street station.

Left: Southern Railway's only Alco RS-11 model also carried the road number 11 and was lettered for Southern subsidiary Carolina and Northwestern (a collection of shortlines in North Carolina and Virginia). Here the rare SR unit, leading three other diesels on train No. 82 between Greensboro and Goldsboro, pulls into the yard at Selma, North Carolina on a warm day in July 1961.

On a hot afternoon in July 1962 a northbound Durham & Southern train rattles across a highway bridge north of Apex, North Carolina where D&S interchanged with the Seaboard Railroad. Baldwin road switcher No. 362 (DRS 4-4-1500) had been obtained from the Soo Line only a short time before this photo was taken.

A newly painted Winston Salem Southbound GP-9 rests between assignments at North Winston yard in July 1962. This 90-mile shortline was owned jointly by N&W and ACL; thus the nearly solid black color scheme was not surprising. WSS's original diesel paint scheme was gray with a red stripe.

Right: A Southern Railway RS-3 lugs an interchange cut upgrade toward the Clinchfield yard in Spartanburg, South Carolina in June 1960. For many years Southern rails provided the only link for the heavy interchange traffic between the Clinchfield Railroad to the north and ACL-subsidiary Charleston & Western Carolina to the south.

Eastbound Tennessee Central freight No. 84, which made an all day run from Nashville to Emory Gap (near Harriman), struggles upgrade near Silver Point (75 miles east of Nashville) behind an FA-1 and three RS-3's in April 1962.

A refurbished front truck is shoved beneath Tennessee Central RS-3 No. 258 at the road's Nashville shops in June 1962.

With sack lunch in hand, the conductor of an eastbound Tennessee Central local climbs aboard his caboose as the train pulls out of Nashville yard on a crisp morning in April 1962.

A veteran engineer on eastbound Tennessee Central freight No. 84 leans out of the cab window of his FA-1 to see his brakeman's hand signals at Monterrey, Tennessee in April 1962.

Servicing outgoing diesels was not an automated process on the "down home" Tennessee Central. Here, on a cool morning in April 1962, FA-1 No. 801 gets a load of sand the old fashioned way - by hand carried pails.

A northbound SAL freight, enroute from Hamlet to Richmond, slams through Apex, North Carolina as two Durham & Southern Baldwin RS-12's switch interchange traffic on a cold afternoon in February 1962.

Acca yard in Richmond was a joint ACL-RF&P facility which also handled all interchange between SAL and RF&P. Here is Acca on a cool afternoon in October 1961 where we find Alco S-2 No. 51 and a four-unit F-7 set, both of which still carried the original RF&P triangular herald.

It was nearly midnight before SAL's premier train, the *Silver Meteor*, paused briefly at Raleigh on its run from Florida to Richmond , from where the RF&P and Pennsy carried it to Penn Station. In August 1962 I made this time exposure of brightly lit rounded-end observation car.

Seaboard hump engines, S-12 No. 1464 and RS-12 No. 1467, head into the classification bowl at Hamlet yard to retrieve an errant cut of cars at dusk on a May afternoon in 1962. Departure yard is in background.

In a view northward from Hamlet yard's hump tower on a May afternoon in 1962, lengthening shadows capture three Geeps and two F-3A's as they pull away from their train and head toward the engine servicing area. A Baldwin switcher will soon be preparing the consist for classification over the hump.

Left: On a warm night in August 1962 I visited the Boylan Avenue tower in downtown Raleigh; this was one of the state's last mechanical interlockers and guarded the crossing of SAL and Norfolk Southern. Setting up a time exposure I first captured the headlights of the southbound *Silver Star* (No. 21) as it swung around a sharp curve (background) and accelerated toward Hamlet. After a short wait I was able to open the shutter again and record a Charlotte-bound NS train, led by two Baldwin road switchers, thumping nosily over the crossing diamonds.

It's late on an August afternoon in 1962 as southbound freight No. 85 from Portsmouth, Virginia pulls into the yard at Raleigh. In the distance a northbound extra waits for clearance.

Its work at Hamlet yard done for the day, a veteran Baldwin switcher (DS-4-4-1000) idles away on a June evening in 1961. Pneumatic tube carried messages between yard offices at either end of the yard complex.

A January 1962 snowstorm is just beginning as northbound SAL freight No. 280 slogs upgrade north of Raleigh behind a GP-9, two sets of FT A-B's and two E-7's. It would take a lot of horses to keep this train moving toward Richmond on this day.

The SAL's Virginia Division included the railroad's busiest mainline (Hamlet to Richmond) as well as the historic line from Norlina to Portsmouth, Virginia. The dispatcher's office in Raleigh was the location of one of the nation's earliest CTC installations during World War II. This photo, made in November 1962, shows a number of additions and modifications to the original machine. Current CTC panels do not include mechanical switches and painted track diagrams such as these. All information is presented via computer screens.

Southbound SAL hotshot freight No. 27 (Potomac Yard to Atlanta) blasts through the village of Apex, North Carolina (south of Raleigh) on a June afternoon in 1962 behind a newly painted RF&P F-3 and five brown and yellow SAL Geeps. Durham & Southern RS-12 No. 1200 switches interchange track.

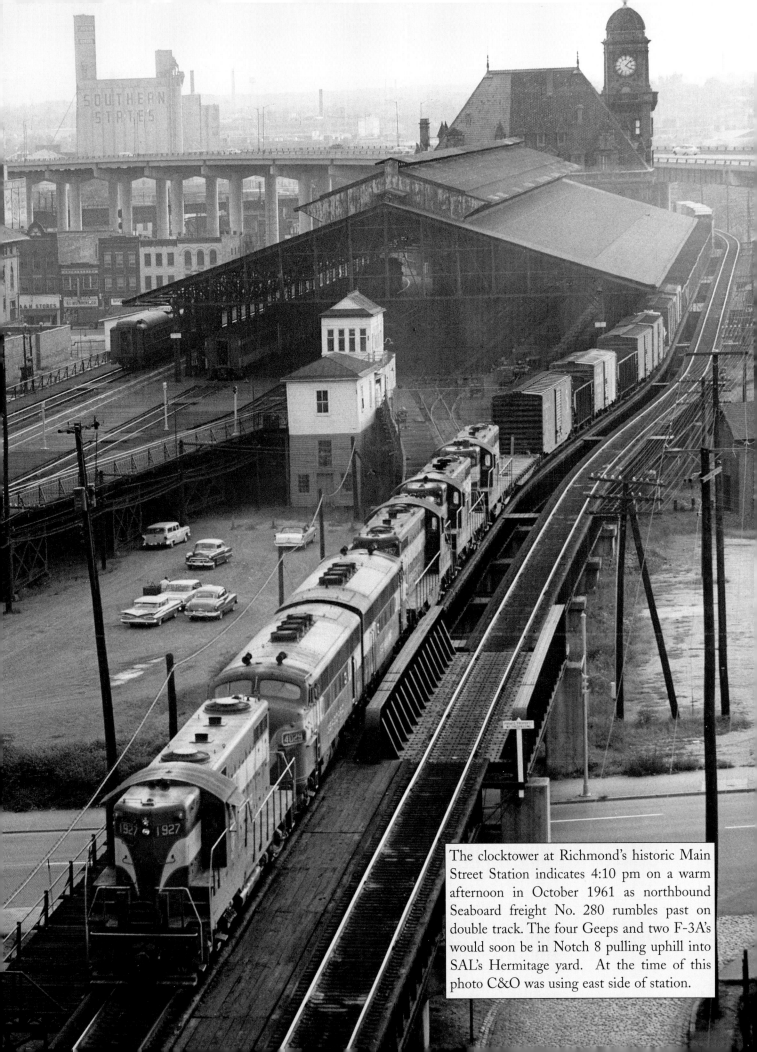

The clocktower at Richmond's historic Main Street Station indicates 4:10 pm on a warm afternoon in October 1961 as northbound Seaboard freight No. 280 rumbles past on double track. The four Geeps and two F-3A's would soon be in Notch 8 pulling uphill into SAL's Hermitage yard. At the time of this photo C&O was using east side of station.

An aging FT unit, built in November 1943 and nearing the end of its career, rests between runs at SAL's Raleigh yard in October 1961. Only a year later this unit would be traded for GP-30's.

Top Right: On a warm night in August 1962 northbound Seaboard train No. 86, led by GP-9 No. 1926 and four EMD mates, awaits the highball from the CTC dispatcher located a half-mile south in an office building.

Bottom Right: On a foggy, drizzly morning in December 1962, the southbound *Palmland* (No. 9), heads into Raleigh behind two E-7's. Green flags indicate a second section would follow.

Three E-7's, two Geeps, and an F-3 power northbound Seaboard freight No. 280 upgrade out of Raleigh on a run from Hamlet to Richmond in April 1962. The train is on a secondary track which has a lower gradient to a summit ten miles to the north.

Seaboard FTA No. 4010, built in November 1943, awaits its highball at Raleigh on a southbound run to Hamlet in August 1961. Baggage cart on adjacent station platform would soon receive a first-class train which must depart before the freight train gets its clearance.

Top Left: On a cold night in October 1962 one of SAL's first passenger diesels (E-4 No. 3008) sits beside RS-3 No. 1540 (originally an RSC-3) and GP-7 No. 1510 at Hermitage yard in Richmond.

Bottom Left: With six routes converging, the North Carolina village of Hamlet was a northern hub of the Seaboard. The classic wooden passenger station stood at the downtown crossing of a north-south double track and east-west line. With a total of 14 daily passenger trains passing through Hamlet in the early 1960's, the L-shaped building was always a busy place. Here is a typical scene on a warm summer night in August 1962.

Above: Southbound SAL passenger train No. 17, the *Tidewater* rests quietly behind E-8 No. 3055 in the Raleigh station after completing its daily run from Portsmouth, Virginia. Before down the entire consist will be turned (without a balloon track) and be ready for a morning departure as No. 18.

SAL Extra 3045 North, an all-TOFC run from Florida to New Jersey, eases past southbound passenger train No. 5 (a Richmond-Atlanta mail express local) at the Aberdeen, North Carolina station on a frosty morning in October 1961. Seaboard's first E-8 (No. 3049) leads the southbound train.

Riding Southeastern Lines

4

In late 1961 I had the opportunity to spend two days riding Seaboard freights between Hamlet and Richmond. Although I had taken on-board photos from Chicago Burlington & Quincy trains while living in Champaign, I was nevertheless quite excited with anticipation as I drove toward Hamlet before dawn. By 6 am I was in the SAL office at the north end of the Hamlet hump yard, and soon found my escort, a Road Foreman of Engines. Our schedule included a ride north on the *Piggyback Special*, a thrice-weekly extra train of trailer-borne farm produce which operated from Florida to the northeast. At the time it was SAL's fastest non-passenger run, covering the 830 miles from Auburndale, Florida (in the heart of a truck farming region) to Richmond in 21 hours. The following day we were to return to Hamlet on manifest No. 75, denoted as the *Merchandiser* on SAL's timecard.

Motive power for the piggyback hotshot was changed out quickly at Hamlet and we were on our way at 7:30 behind E-7 No. 3045 and two Geeps. In order to have more room to move about and get additional shooting angles, I decided to stay in the third unit until we arrived in Raleigh for a crew change. With 2.5 horsepower per ton the engineer was able to hold the short consist to passenger train speeds over most of the run. Soon after leaving Hamlet we were zipping downhill at 65 mph toward Aberdeen where

Southbound RF&P passenger train No. 93 heads into Richmond on a bright day in October 1961. This view is from the cab of a Seaboard switcher delivering a train of piggyback trailers from SAL's Hermitage yard to the ACL-RF&P Acca yard. The RF&P F-units ahead stand ready to highball northward with our train.

On a gloomy day in October 1961 the engineer of southbound Seaboard freight No. 75 pumps air on a 35-car train at Richmond's Hermitage yard as the northbound *Silver Star* (No. 22) zips past behind a pair of RF&P E-8's on a run to Washington's Union Station.

we found southbound mail and express train No. 5 (bound for Atlanta and Birmingham) working at the station. Our speed dropped to 35 as the *Special* struggled up the 1.1 percent hill between Aberdeen and Southern Pines. The Road Foreman noted that this was the ruling grade between Hamlet and Richmond.

At Vass siding we had a running meet with No. 9, the *Palmland,* and then at Sanford eased into the siding and crept past freight No. 89, the *Tar Heel.* Neither train was required to come to a complete stop as the dispatcher in Raleigh once again performed electronic magic with track switches and signals. As a result of these deft mainline maneuvers, we covered the 92 miles into Raleigh in an hour and 45 minutes, about the same time as Seaboard's premier passenger train, the *Silver Star*. As we passed between the platforms of the Raleigh station I noticed two trains waiting in the adjacent yard for us to clear. No. 61 was a Raleigh to Hamlet local while No. 82, led by three Geeps, was heading for Portsmouth, Virginia.

After a quick crew change at the north end of the yard we received a green signal and the engineer notched up Extra 3045 North for its final leg into Richmond. Once on the main we were able to move along smartly, slowing only for a few wayside villages. A yellow signal at Manson led us up the "wrong main" on an eight-mile stretch of double track into Norlina in order to clear a work train dumping ties. Speeding northward into Virginia we topped the LaCrosse summit and were then on a generally descending grade into Richmond. The dispatcher set up another meet at Alberta where manifest No. 75 was in the hole, and soon we were cruising down a twenty-mile stretch of straight and level track known to the crews as the "drag strip" which allowed our speed to hold at 65 mph.

Between Petersburg and Richmond the lines of Seaboard and ACL played "crossover" with each other. At Lynch, four miles north of Petersburg, the Coast Line vaulted overhead while six miles later, at Chester, the SAL had the upper route. In Richmond we eased across the James River and then passed the venerable Main Street station and its magnificent clock tower. Soon we were pulling slowly out of the river valley on

Seaboard freight No. 85, headed for Norlina and Raleigh, roars past the classic wooden station at Suffolk, Virginia on a cold overcast day in February 1962. Immediately behind the four Geeps were two wide loads (boats) and an Army battle tank.

a one percent grade into SAL's Hermitage yard. Extra 3045 North finally eased to a halt at 12:30 pm after covering the 254 miles from Hamlet at an average speed of 55 mph, an impressive performance for a sprint train in 1961. As soon as the road power was cut off, a Baldwin switcher moved in to deliver the consist to RF&P's Acca yard. Reluctant to let the trip end, I climbed aboard and continued with the TOFC flats as they were dropped off beside a trio of blue and gray F-7's which would soon be speeding toward Potomac yard with both the Seaboard cars and a similar piggyback cut from ACL.

At 7:45 the next morning I was eager for another operating day as I boarded a six-unit consist (two GP-9's, and FT A-B set, and two GP-7's) at the Hermitage ready track. Today's run would take considerably longer than did yesterday's piggyback sprint. With No. 75's schedule the SAL was willing to sacrifice road speed in order to expedite key interchange pickups. After we moved the locos down to the train and began pumping air, I noticed a headlight of a northbound train through the fog. It turned out to be No. 22, the northbound *Silver Star,* which had traded its white Seaboard units for a pair of blue and gray RF&P E-8's at Broad Street station. At 8:55 the short SAL train eased out of Hermitage yard and coasted downhill into C&O's Brown Street yard for the first pickup which swelled our consist to an even hundred cars. Then it was back to the mainline for 28 miles into Ryan and a 23-car N&W pickup which gave our train nearly 6000 tons. At a few minutes past noon No. 75 eased to a stop at Norlina to add 27 cars of new autos and other merchandise off the Portsmouth line. From here it was a straight shot into Raleigh.

The new crew which came on board at Raleigh anticipated a nonstop run to Hamlet but it was not to be on this day. Drag freight No. 88 was running late because of sticking brakes on its whopping 180-car train which was too long for any of the sidings between Hamlet and Raleigh. Thus the dispatcher was forced to split No. 75 with its 150 cars into two cuts at the New Hill siding, 21 miles south of Raleigh. The crew, much to their disgust, had to cut the train 100 cars back for a road crossing and then slide the locos plus a few cars into a short siding on the opposite side of the main line. After an hour's wait we finally spotted No. 88's headlight on the horizon. It then disappeared and reappeared twice as the train dipped into sags on the rolling terrain. By 4:30 the crew had reassembled No. 75 and was heading onto the main with a clear route to Hamlet, 72 miles away. Soon however, we hit a blinding rainstorm that cut visibility to almost zero. After passing through the village of Moncure we were soon grinding upgrade toward Southern Pines on wet rail which produced a constant blinking of the wheel-slip indicator light on the control panel. Fortunately, the rain subsided before we reached Aberdeen and we were able to make good time into Hamlet. At 6:15 the engineer eased the mammoth manifest to a halt at the north yard office in Hamlet and my eleven-hour trip had ended. But what an unforgettable experience these last two days had been.

A few months after these Hamlet-Richmond runs I made another two-day trip, this time a round trip between Raleigh and Portsmouth on SAL freights Nos. 82 and 85. Each of these runs was assigned a different job. No. 82's role was to do local work with special emphasis on the pulpwood and paper industries around Roanoke Rapids, North Carolina and Franklin, Virginia. Conversely, No. 85's primary responsibility was to hustle autos and other expedited traffic from Portsmouth to Norlina for pickup by No. 75. My trip on No. 82 began as I boarded the third unit (GP-7) of a three Geep consist at 8:30 on a chilly February morning in 1962. We headed to the north end of the yard and coupled to our train, composed of 22 loads and 28 empties (2268 tons). As we were crawling upgrade out of the Raleigh yard, the fireman had to scramble out on the walkway, open a couple of hood doors and spend some time trouble-shooting the last unit because it would not load up when the throttle was higher than Notch 6. However, by the time we finished our first switching stop at Neuse (10 miles out) it was working properly. At Neuse we set out some empty wood racks and replaced them with loaded ones. I learned that crews often referred to the ubiquitous logs as "North Carolina bananas" because they had to be delivered to the paper mills on a regular and reliable basis.

There were additional switching stops at Youngsville, Franklinton, and finally Henderson, where we again exchanged empty racks for loads. We left Henderson at noon with 55 cars and, after a quick

The fireman on northbound SAL freight No. 82, headed to Portsmouth, Virginia, checks inside the hood of the third Geep as the train speeds toward Wake Forest, North Carolina on a bright winter day in February 1962. After leaving the Raleigh yard, the engineer found that the unit would not "load up" electrically the proper way but, after some tinkering, the crew was able to get all the switches and relays inside the veteran Geep's control system to work satisfactorily.

SAL train No. 18 for Portsmouth, pulls into the fog-shrouded station at Franklin, Virginia just after dawn on a February day in 1962 while southbound freight No. 85, headed toward Norlina and Raleigh behind four Geeps, swings past on the siding.

pickup stop at Manson, arrived at the south end of Norlina yard to find our companion train No. 85 shoving a transfer cut into a lead track for the waiting manifest No. 75. After No. 85 left for Raleigh we were able to pull into the clear at 1:05 pm. The crew decided to take a short lunch break at a nearby cafe to await the arrival of northbound manifest No. 280, from which we were to get 28 auto racks. By 2:30 the crew had reassembled their train, and we left for Roanoke Rapids where we set off most of our loaded wood cars and picked up boxcars loaded with paper products. Leaving at 4 pm we ducked under the ACL mainline at Weldon as a cold rain began to fall through the gathering dusk. At 5:15 we pulled into Hand siding to meet Raleigh-bound passenger train No. 17, the *Tidewater*, which connected with the *Silver Comet* in both directions. Our remaining log cars were set out at Franklin paper mills. It was ten minutes past 6 pm and completely dark when we cleared Franklin. Our final stop was at the N&W interchange at Kilby where, as we were picking up a couple of tank cars, a loaded coal train rumbled past on an elevated line. We finally pulled to a stop in the Portsmouth yard at 8:10 after almost a dozen hours on the road.

It was still dark as I got to the yard office at 6 the next morning. No. 85's train was already assembled and waiting for us to board. The consist included 50 loads and 37 empties but, with 27 loaded auto racks, had an equivalent length of almost 100 cars and a tonnage of 3655. There were 48 cars, including the autos, to be transferred to No. 75 at Norlina. Due to the extra tonnage a fourth Geep had been added to the three used on No. 82 the day before. Leaving at 6:30 in the first light of dawn, we crawled cautiously through a thick, coastal fog, passing a work train on the siding at Carrsville and later easing past the eastbound *Tidewater* (No. 18) at the Franklin station. Our first switching stop was Roanoke Rapids to pick up 22 empty wood racks. We left at 10:15 for a nonstop run to Norlina where, an hour later, we set out the cut for No. 75 and picked up more empty wood cars. As on the previous day, the crew was allowed a short lunch break while No. 75 was being reassembled. On this day the dispatcher was kind to No. 85 for, after leaving Norlina at ten minutes before 1 pm, we had only one meet (No. 82 at Greystone siding), and then slid past hotshot freight No. 280 in the Henderson yard and mail-express train No. 4 on the double track stretch around Kittrell. Thus our arrival in Raleigh was at 2 pm, a relatively quick run as compared with the previous day's No. 82.

These four trips on the Seaboard allowed me to see up close how exacting and difficult it is to keep tonnage and passengers moving across a division which includes both high speed pass-through movements and locals serving congested industrial regions. But I must confess that, for a few weeks after these trips, I was on a "high" from the intensity of the experience of having logged almost 900 miles on one of the South's busiest mainline districts.

In June 1962 I had a chance to see another facet of the southeastern railroad picture when I trekked westward to the Volunteer State to catch a ride on Tennessee Central freight No. 84 from Nashville to Emory Gap. My trip began at 6:30 am at the downtown river front yard. Our train was led by FA-1 No. 801 and two RS-3's. Behind the three Alco's were 74 cars at 2834 tons. The turbocharged diesels had to work hard going upgrade on Donelson hill but soon ground to a stop at Stone River siding for a meet with counterpart train No. 81 from Emory Gap which had backed onto the Old Hickory branch because the siding was too short. TC covered its eastern mainline with two crews on trains 81 and 84. A Nashville crew worked to Monterey (108 miles) and were relieved by a second crew which worked a turn run to Emory Gap (55 miles each way). By the time No. 81 arrived back in Monterey and the train was classified, the Nashville crew was rested and available to complete the 26-hour round trip.

After the meet at Stone River was completed, No. 84 worked slowly eastward through lush mountainous country until the first stop at Martha (23 miles out) where we set off 17 empty hoppers on a passing track, then backed around them and pushed the cut uphill into a cement plant. Later at Lebanon the crew gave a hearty greeting to the operator who came out to check the train as we passed. Beyond Buffalo Valley (MP 70) the Alco's began to dig into the Silver Point hill (eastbound ruling grade) and, as the gradient increased to 3 percent, our speed dropped steadily. The throbbing engines, shrouded in clouds from the sanding pipes, were eventually reduced to a 5 mph crawl before reaching the summit just west of the tiny Silver Point depot. Seven miles later we stopped at Baxter (named for the father of the Tennessee Central) to set off a car of stone and pick up an empty boxcar. Pulling into Cookeville at 11 am I noticed that, at the station, there was another monument to Mr. Baxter in the form of a stone monolith with a metal plaque.

The crew took a short lunch break at Cookeville and then worked their way through an extended series of switching moves before heading out at 12:30 as a summer shower began to fall. After passing through Brotherton, 10 miles from Cookeville, our train faced another formidable grade, Algood Mountain. With the rain now in torrents, the train slogged through rock cuts, negotiated sharp curves with flanges squealing, and bored through stretches of track where rain soaked tree limbs scrapped against the cab sides. Wheel slip lights blinked continuously while the brakeman, who sat with me in the third unit, worked the sanding valve to help the RS-3 get more traction on wet rail. Our speed dropped to 7 mph, then 6, 5 and finally only 4 mph on the twisting climb around the mountain. After a half-hour the rain tapered off and finally stopped, leaving the mountainside steaming with evaporation. Once we crested the hill the engineer was able to get our speed back to a 30 mph cruise before slowing for Monterey at ten minutes until 2 pm. In the yard the Nashville crew uncoupled their caboose and then cut the train for a road crossing before heading back to the end of the train to retrieve the crummy and shove it into a spur. Their day ended at 2:30 as the three weary Alco's were halted in the service area.

A half-hour later the Monterey crew came on duty and began to assemble their train. They first detached the trailing RS-3 (No. 257) for switching work while the other units were being serviced. Later, with all units together once again, they pulled a cut of loaded coal hoppers onto the Wilder branch for weighing. Finally at 5:20, with another rainstorm brewing, No. 84 chugged slowly out of Monterey with 69 cars (including 40 coal-laden hoppers) and a tonnage of 4392, quite a load for three Alcos's on a mountain rail-

On a warm June day in 1962 the conductor of Tennessee Central train No. 84 confers with the agent-operator at Baxter, Tennessee about some local switching duties. An Alco FA-1 and two RS-3's provide power; this was the view from the last unit.

road. Two miles out of Monterey at Clinchfield siding, we passed a mine run behind RS-3 No. 252 and, averaging about 25 mph, rolled toward Crossville, where we stopped to pick up 10 cars at a pulpwood yard. Leaving Crossville, we were soon into the heart of the Cumberland Mountains of eastern Tennessee. The trackage leaped over numerous ravines on high, timber trestles and steel bridges, many of which were on curved segments or even S-curves. It was nearly dark as we eased to a stop in the village of Crab Orchard, which was surrounded by timbered peaks. We set out five cars for a limestone quarry which seemed to be in the midst of removing the side of a nearby hill. After passing Daysville at MP 149 we began our descent from the Cumberland plateau. On the sharply curved Piney Creek bridge, while flanges squealed in protest, the FA's headlight beam swept across a nearby rocky cliff face until it finally pointed directly at a tunnel portal.

Our train emerged from the bore surrounded by thick acrid smoke from brake shoes which held the consist in check on a 2.6 percent grade (ruling westbound gradient). The line clung to a mountainside ledge, and spread out before us was a panoramic view of the industrialized valley between Rockwood and Emory Gap which was surrealistically illuminated by the flickering orange flames from two steel mills. We stopped at the mill in Rockwood at 7:45 to set out four hoppers of coal, and later ran parallel to the CNO&TP mainline on our final leg into the Emory Gap yard which was on a slight upgrade. Pulling into the yard at 8:15, the crew quickly uncoupled from the train and pulled the locos into the service area where the last RS-3 was replaced by a companion unit. The power consist was wyed and then headed back to the front of an already-classified No. 81. By 8:45 the westbound train was ready to leave on its overnight run to the Tennessee capital. And I, after fourteen hours on the road, was ready for shower, a hearty meal, and a soft bed. This had been no ordinary "cab ride," like the earlier ones on the SAL, but a real adventure whose memories will linger as long as I live.

Eastbound Tennessee Central train No. 84 meets its westbound counterpart No. 81 at Stone River siding, ten miles east of Nashville, in June 1962. Each train is powered by an FA and two RS-3's.

Accelerating to mainline speed, the *Southerner* splits the semaphores south of the Meridian yard as it begins its last lap into New Orleans on a May afternoon in 1966. Open MU-receptacle on the nose of this black and white E-8 is evidence of the Southern's practice of removing a third unit at Birmingham. Number on signal mast indicates 4.6 miles to Meridian, the beginning point of the New Orleans and Northeastern Railroad.

End of an Era

5

After eight years of apartment living at four addresses in three states, we settled into our own house in Austin during January 1963. One of my wife's first requests was, "Let's stay here at least five years. I'm tired of moving." I suppose one could say that I have agreed with her on that score. We are still at the same address, and have remodeled our house recently with a view toward retirement. (A major part of the remodeling was the installation of a darkroom, a first for me.)

In contrast to the constant railroad activity around the Raleigh area which, in retrospect, provided a distraction for my professional advancement, there was virtually no temptation in the immediate vicinity of Austin. Unlike Raleigh, the capital of the Lone Star state was merely a pass-through point for both MKT and Missouri Pacific. I learned quickly that, in Texas, one usually had to travel much farther to find trains and yards than was typical in the southeast. Thus my camera stayed on the shelf except for occasional weekends and during the breaks between school terms when we returned to Mississippi for visits.

During my first few years in Texas I began to accept research positions in government or industrial labs during the summers. During the 1960's I spent parts of three summers at an Air Force facility near Tullahoma, Tennessee which is situated on the former NC&StL mainline between Nashville and Chattanooga. In the long summer afternoons after work I often drove about ten miles north of Tullahoma where there was a shady spot beside a country road overpass on the L&N mainline, and just waited for a train to appear. It was so quiet among the farms in these rolling hills that, between the birds chirping and the farm tractors chugging in the distance, the highly trained ear of a railfan could pick up the sound of a diesel horn nearly five miles away. The L&N often ran both coal and TOFC hotshots in late afternoon as well as locals heading home to tie up.

If an especially photogenic power lashup passed on a southbound train it was easy to pace the train all the way to Cowan (18 miles from Tullahoma) which lay at the foot of Cumberland Mountain. The steep climb to the summit tunnel required NC&StL and later L&N to keep one or more helper sets at Cowan. Usually these were coupled to the rear at Cowan yard and pushed southbound trains to the summit where they were cutoff on the fly. For northbound trains the helpers were attached at Sherwood.

Whenever I could get away on weekends I would head north to catch action around L&N's Radnor Yard in Nashville or a few miles further north to the Shelby Park bridge over the Cumberland River. Other times I drove south to Chattanooga and watched Southern push a steady stream of trains across the high bridge which spans the Tennessee River near TVA's Chickamauga Dam. Just as Radnor yard was a centrally located L&N hub for five diverging routes, Southern's Citico Yard (now named for former president Harry DeButts) sat at the confluence of an equal number of lines. The Chickamauga bridge, incidentally, was originally double tracked but, with the introduction of CTC on the CNO&TP, was single tracked and then, with rising traffic density, was returned to double track with bidirectional signaling.

At the conclusion of my summer work in 1965, I drove to the eastern Tennessee town of Johnson City to meet Wiley. Along the way I again paced Tennessee Central freight No. 84 which was no longer led by an FA but by a recently purchased, low-nose Alco RS-36. Wiley and I spent a couple of days chasing Clinchfield trains out of nearby Erwin yard and enjoying the mountain scenery surrounding both Southern and Clinchfield trackage. At this time the CC&O was experiencing a power shortage and most of the trains carried either leased units, such as RF&P F-7's, or one of Clinchfield's switchers in their consists. The highlight of the trip was a chance to pace a northbound merchandise train out of Erwin behind F-7 No. 800, the road's initial cab unit (December 1948) which had been rebuilt from an F-3 in 1952.

With most southeastern roads having embraced diesel power immediately after World War II, one would naturally expect that they would be ready to invest in new locomotive designs when their classic diesel units came to the end of their prime service lives

Above: Wreathed in brake-shoe smoke, the northbound *Dixie Flyer* eases to a stop at the classic depot in Cowan, Tennessee after cresting the Cumberland Mountain barrier and its summit tunnel in June 1964. Lead Geep carries the original L&N diesel scheme while the second unit is in "stealth gray."

Top Left: A trio of cab units lead a long freight southward from Cowan, Tennessee in late afternoon of a warm July day in 1965. With throttles in Notch 8, the F's are beginning the steep approach to Cumberland Mountain tunnel. The two lead units were purchased new by the L&N while the third came from the NC&StL as did this mainline between Nashville and Atlanta. The substantial grades over this summit required pushers in each direction; on this train it was a trio of ex-NC&StL Geeps.

Bottom Left: In August 1965 a Chicago-Atlanta TOTE train, running as a second section of No. 95, the *Dixie Flyer,* slices through the Tennessee hills north of Tullahoma behind an A-B set of F's. Lead unit has an inactive steam generator and is classified by C&EI as an F-7 while the second unit is an L&N F-3.

The shadows are lengthening in late afternoon as a Salisbury-bound train whines downgrade on dynamic brakes between Ridgecrest and Old Fort, North Carolina in August 1973. These four F-7's were among the last of their breed on the Southern Railway and the last SR cab units I ever saw in freight service.

after fifteen years. Although Alco had long used the supercharger as a way of boosting power, EMD had stuck with normally aspirated engines until 1959 when it introduced two models with "souped up" turbocharged engines. The GP-20 was a natural extension of the GP-18 while the SD-24 was a more powerful descendent of the SD-18. Even though the GP-20 was sold extensively to western roads, the only eastern line to buy any was New York Central (which got fifteen). With an economic recession in progress the GP-20 model sold only 260 copies, while the SD-24 was even less successful with but 224 produced. The only southeastern road to buy SD-24's was the Southern Railway which acquired twenty-five in 1959-60 to expedite tonnage on the busy and mountainous CNO&TP line.

However, these two new EMD models are significant in the evolution of diesel power development in the U.S. because they represented the initial attempts to develop a "second generation" locomotive, a process which gained momentum with the unveiling of EMD's GP-30 model in October 1961. Encouraged by a new GM locomotive trade-in policy, most large southeastern roads began replacing their aging fleets of cab units in the early 1960's. Of the early buyers of GP-30's in Dixie, only the ACL (with ten units in January 1963) failed to transition quickly to this unit for mainline freights. At the other extreme, GM&O bought thirty-one copies using its FA/FB models as trade-ins, Seaboard acquired thirty-five, L&N got fifty-eight, and the Southern Railway's high-hood fleet eventually totaled 120 units. Needless to say these 254 GP-30's, which represented over a quarter of the total production of 948, quickly decimated the ranks of classic diesel models in the southeast during the early 1960's.

Thus each time I returned to the region during this period, I would see fewer and fewer GM&O FA/FB's or Southern F-7's at work. Three years after moving to Texas I was visiting Meridian in late summer and happened across a GM&O local heading north one morning on its all day run to Okolona, Mississippi. Leading the train was an RS-2 followed by a faded and tattered FA No. 717. I assumed that the same units would return to Meridian the next day with the 717 in the lead, and so I ventured northward the next afternoon to intercept the train. Alas it was not to be; the units had been wyed at Okolona and the FA was still trailing, probably because of its antiquated brake system and related control equipment. As it turned out this was one of the last operating FA's on the entire GM&O. Most had already been traded for GP-30's and -35's. Number 717 would soon be dismembered by the scrapper's torch.

During my first decade in Austin, there were many times when professional travel would take me to the eastern seaboard. Whenever possible I would stop a few days in Raleigh and visit with Wiley Bryan so that we could continue exploring SAL (and later SCL) lines as well as the scenic Clinchfield route between Marion, North Carolina and Kingsport, Tennessee or the Southern Railway lines around Asheville. It was on one of these trips in 1973 that I photographed Southern F-units for the last time. Early one morning we shot a westbound freight going uphill toward the Ridgecrest summit behind a lead GP-38 and four F-7's (A-B-B-A). Fortunately, upon arrival in Asheville this power consist was sent back to Salisbury that afternoon with the classic F's in the lead. Thus, just before the sun dipped behind the Blue Ridge mountains, we were able to capture a few shots as the train eased downgrade between the Andrews Geyser loop and the town of Old Fort.

Among the final first generation diesels working in Meridian, aside from the M&B's GP-7 and -9 units (one of which is still active in a rebuilt state), were a pair of ex-Illinois Terminal RS-1's (still in their IT colors) used as yard power in the early months of the Illinois Central Gulf era. But the most elegant of the classic units to pass through Meridian in the 1970's were the refurbished E-8's used on the *Southern Crescent*. Southern's fleet of E-8's (all A-units) totaled seventeen, the first seven of which were delivered in September and October of 1950 and numbered 2923-2929. Three years later ten more were built and lettered for the NO&NE with numbers 6906-6915. The last unit of this order, completed in December 1953, represented the final E-8 assembled in LaGrange. As built, these lean racers were adorned with Southern's traditional green with white and gold trim. However, beginning with the SD-24's in 1959, the green was replaced by black on all units, and the tradition circular herald was deleted from the noses of F-units. The E-8's, in their black and white incarnation, had minia-

Accelerating out of Meridian on a crisp December day in 1973, the northbound *Southern Crescent* heels to a superelevated curve four miles north of the station behind a quartet of classic diesels, green and white E-8's with the train's name on their noses. The extra-wide right-of-way once carried the double-tracked Alabama Great Southern mainline from Meridian to Bessemer, Alabama.

ture decal heralds fastened to their noses.

In the first years of the *Southerner* operation just after World War II, the road would reduce power to one unit south of Birmingham since the terrain was relatively flat and the consist light. During the 1950's and 1960's two units became standard for a 10 to 12 car train but, near the end of its operating history, the *Southerner* again carried but one unit and usually less than a 6-car train through Meridian. With the coming of Amtrak in 1971, Southern's council of executive vice-presidents recommended initially to President Graham Claytor that the road sign a contract for operation of Amtrak trains. But Claytor, normally a strong consensus builder with his management team, decided ultimately that the Southern Railway System, would *not* participate, and thus it became the largest U. S. railroad to stay "independent." (D&RGW was the other holdout.) Instead Claytor launched in mid-1972 one of the nation's last two privately-operated long-haul passenger trains, the thrice-weekly *Southern Crescent* which ran on the same route as the *Southerner* and connected with Amtrak's *Sunset Limited.* Northeast connections at Washington Union Station enabled one to ride on a sleeping car all the way from Boston to Los Angeles via Atlanta and New Orleans.

To handle this first-class train, Southern decided to return its E-8 fleet to their original paint scheme and even added the train's name in gold-tinted, script lettering below the front number boards on each unit. During their rebuilding the seventeen units received a coherent number sequence in which Nos. 2923-28 became 6900-05 while No. 2929 became 6916. It was a nostalgic treat to watch (and photograph) three of four of these beautiful machines, the icons of southeastern dieselizaton, as they accelerated the *Crescent* out of Meridian on its final leg to New Orleans. During the 1970's one of my favorite spots to shoot this train (and many others) was from the grounds of a U. S. Fish Hatchery which was situated adjacent to the mainline a few miles south of Meridian. The front lawn, which touched the tracks, was always mowed neatly and there were nearby trees with low-hanging limbs, perfect for framing speeding trains in either direction.

During the bicentennial celebrations, Southern decorated these classic units with the names of southern colonialists who signed the Declaration of Independence. For example, No. 6900 carried the

A six-unit mixture of cab units and Geeps power northbound Clinchfield merchandiser No. 97 through Johnson City, Tennessee on a cloudy afternoon in August 1965. Train is enroute from Erwin to Elkhorn City, Kentucky where it will be handed off to the C&O. Lead unit was CC&O's first cab unit, built as an F-3 in 1948, and converted to an F-7 in 1952. It was later saved from the scrapper's torch and converted to special executive passenger service.

name of *Benjamin Harrison* of Virginia. My final photos of the SR-operated *Crescent* in Meridian were made during a Christmas visit in 1978. By then the once resplendent paint jobs on the locos were showing considerable weathering. When Southern finally relented and allowed Amtrak to operate on its lines, the railroad began to purge these units from its roster. Two of the classic locos were given to preservation groups (No. 6900 to Historic Spencer Shops in North Carolina and 6901 to the Atlanta Chapter of the National Railway Historical Society). Another pair were wrecked in early 1978 (Nos. 6906 and 6910) while two others (6902 and 6915) were traded to EMD in April 1979. A half-dozen more were sold to the New Jersey Department of Transportation and continued in service while five others were sold to Precision National Corporation.

Blue and white Alco S-1 No. 321 switches a lumber yard at Fulton, northern terminus of the 24-mile Mississippian Railway in August 1976. A pair of S-1's replaced two 2-8-0's on this shortline in 1967. The two vintage Alco's were built for the Erie Railroad and continued to carry their original numbers (314 and 321) through numerous owners.

An A-B-B-A set of nearly new Alco cab units (FA/FB-2) eases out of DeCoursey yard near Covington, Kentucky with freight No. 43 on a warm August day in 1956. Lead unit would eventually become L&N's last active FA.

A northbound train growls slowly out of Nashville's Radnor yard on an overcast day in May 1967. Lead FA-2 No. 368 (built February 1952) is assisted by a pair of F-7's and a Geep.

Late afternoon illumination highlights the running gear of a trio of black F-7's as they ease a southbound train from Louisville across the Cumberland River in Nashville in July 1964.

Below: A Tennessee Central train from Nashville to Hopkinsville, Kentucky has stopped in Clarksville, Tennessee to do some switching on a hot September day in 1959. Meanwhile, FP-7 No. 911, heading a northbound L&N local, is also engaged in setouts and pickups on the lower tracks.

Above Left: A cool April morning in 1962 finds a Southern Railway train from Knoxville to Asheville snaking through a fertile valley which surrounds the French Broad River near Newport, Tennessee. Five EMD cabs were needed to maintain track speed on this curving, hilly line in the picturesque Smokey Mountains.

Above Left: The northbound *Southern Crescent*, led by a trio of E-8's, speeds toward Meridian. In this August 1972 scene the lead unit is still in the Brosnan-era black and white paint but the other two units have already received the new green colors which would adorn the final fleet of Southern Railway passenger locomotives.

Above: On a warm afternoon in July 1965 the southbound *Royal Palm* rumbles across the Tennessee River at Chattanooga behind an E-7/FP-7 duo. The *Palm,* once a major entry in the Midwest-Florida streamliner sweepstakes, was eventually reduced to a largely headend train as this photo suggests.

Brightly painted caboose and GP-9 display two variations of Meridian & Bigbee heralds at Meridian in August 1958. Colors are dark blue with yellow trim.

Top Left: On a hot August afternoon in 1957 a Meridian & Bigbee train from Myrtlewood, Alabama pulls past the GM&O freight house into the Meridian yard with transfer cars for the Illinois Central and GM&O. A close inspection reveals a horse-drawn wagon above the second unit (a GP-7) which carries number 1 (later 101) because it was M&B's first diesel. Lead unit (102), a GP-9, was the second purchase by the shortline. Serving a large paper mill in Alabama, M&B's outbound traffic included full boxcars and empty pulpwood racks while inbound the boxcars were empty and the racks full of logs.

Above: A pair of black and white RS-2's head an eastbound Atlantic and Danville train near Emporia, Virginia in May 1961. In less than two years the independent (but bankrupt) A&D would be sold at auction and become a part of the N&W system.

The Mobile and Ohio Railroad constructed this cathedral-like station and office building in its namesake city in 1907. It was used by successor lines GM&O and ICG until 1986. A low, winter sun in December 1985 highlights the exquisite detail of one of the South's most elegant stations.

Black and white Mississippi Export Railroad NW-5 classifies a cut of cars at Moss Point, site of the road's offices and shops in August 1965. This rare road switcher model, a forerunner of the BL-2, was built in January 1947 for the Fort Street (Detroit) Union Depot.

Like birds flocking to their roost, the Meridian terminal after dark was usually filled with resting units whose idling diesels filled the night with a dull hum. On a warm August evening in 1962 a group of green-clad Southern Railway Geeps and an EMD switcher await the dawn of another work day.

After switching the N&W interchange track at Denniston, Virginia an eastbound Norfolk Franklin & Danville local heads out for LaCrosse on a sunny afternoon in May 1968. The formerly independent Atlantic & Danville Railroad was renamed NF&D and became a subsidiary of the N&W in October 1962. Hence both motive power and caboose were property of the parent company.

Epilogue

6

While assembling the photographs for this book I was struck once again by how few of the South's early diesels were saved, even as cosmetically restored carbodies. Fortunately, there are a few examples of the southeast's first generation machines still around to admire - including the first production FT (ex-Southern), an Atlantic Coast Line E-3, and a Columbus & Greenville DRS-6-4-1500. Moreover, the general attitude throughout the nation about preserving the physical evidence of our earlier technology has become much more positive (with spectacular results) during the last fifteen years. This bodes well for the future, but we are a long way from catching up with our European counterparts when it comes to railway preservation.

Just as the motive power pictured on these pages has disappeared, so too have most of the railroad names which were displayed on the sides of these classic diesel units. Of all the Class I lines whose early motive power is pictured here, only the Illinois Central name is still in use, even though the railroad itself has changed considerably. Interestingly, the diminutive RF&P was one of the last to disappear, having only recently been swallowed up by CSX. Fundamental changes have also occurred in many of the cities and towns mentioned in the text. Although the crossing diamond at Opelika still shudders beneath hefty tonnage levels, the locos carry banners of the two southeastern giants, CSX and Norfolk Southern.

Meridian also remains an interesting place to watch trains of NS, Kansas City Southern, and the still independent M&B. Indeed, the Meridian railroad scene has been transformed dramatically by KCS's recent thrust into the southeast. Working closely with Norfolk Southern, KCS has transformed the persistently underutilized Meridian-Shreveport line into the key segment of a new high-speed transcontinental route. On the other hand, one can only lament the demise of the historic M&O mainline from Mobile to Jackson, Tennessee. Although no longer a continuous route, the segment between Meridian and Tupelo remains as a busy KCS artery.

But perhaps the saddest place for me to contemplate is Raleigh, which has seen the once-bustling SAL Virginia Division mainline become a collection of disconnected branch lines on the CSX. The new Norfolk Southern Corporation is now the major player on the scene, having inherited the *old* Norfolk Southern Railway as well as the ex-Southern Railway lines in the area. Of course, no one was sadder about this evolution than the late Wiley Bryan who, during his last decade, would see his "railroad roots" around Durham and Raleigh gradually disappear.

In June 1962 a westbound N&W freight rumbles past the classic wooden depot at Blue Ridge, Virginia, 12 miles east of Roanoke.

Shaffer's Crossing engine terminal, one of the last bastions of steam power, finally succumbed to the diesel in the late 1950's. Here we see a batch of black GP-9's along with an Alco T-6 on an April morning in 1962. In the right background is a now obsolete Y6b 2-8-8-2 while the water tank and its downspouts symbolize the earlier steam power era.

Norfolk & Western's eastbound *Pocahontas* (No. 4) speeds past the classic wooden depot at Blue Ridge, Virginia behind a pair of leased RF&P E-8's in June 1962.

In June 1962 an ex-VGN Trainmaster, in faded black and yellow colors, is framed between a Geep and a notched-nose DL-701 at Shaffer's Crossing engine terminal in Roanoke. Some of the N&W Trainmasters were traded for Alco C-630's which utilized the FM trucks.

In August 1973 the caboose of southbound Clinchfield merchandiser No. 94 has just cleared the short Washburn tunnel No. 2 south of Altapass, North Carolina on its run to Spartanburg where the train will be interchanged with ACL subsidiary Charleston and Western Carolina for delivery to the ACL mainline at Yemassee, South Carolina.